Football, Gambling, and Mc

Fausto Martin De Sanctis

Football, Gambling, and Money Laundering

A Global Criminal Justice Perspective

Springer

Fausto Martin De Sanctis
Tribunal Regional Federal 3rd Region
Sao Paulo
Brazil

ISBN 978-3-319-05608-1   ISBN 978-3-319-05609-8 (eBook)
DOI 10.1007/978-3-319-05609-8
Springer Cham Heidelberg New York Dordrecht London

Library of Congress Control Number: 2014936297

Springer is part of Springer Science+Business Media (www.springer.com)

Acknowledgments

This book was possible thanks to research completed at the U.S. Law Library of Congress in 2013. I want to express my gratitude for having the opportunity to access documents, papers, and books, and I thank all of the people at the Law Library of Congress, especially Eduardo Soares, for their invaluable help. I am also grateful to Rebecca Szucs for her excellent work editing this book.

Contents

About the Author

Fausto Martin De Sanctis holds a Doctorate in Criminal Law from the University of São Paulo's School of Law (USP) and an advanced degree in Civil Procedure from the Federal University of Brasilia (UnB) in Brazil. He was a Public Defender in São Paulo from 1989–1990, and a State Court Judge, also in São Paulo, from 1990–1991, until being appointed to the Federal Courts. He is currently a Federal Appellate Judge in Brazil's Federal Court for Region 3, with jurisdiction over the states of São Paulo and Mato Grosso do Sul.

Judge De Sanctis was selected to handle a specialized federal court created in Brazil to exclusively hear complex cases involving financial crimes and money laundering offenses. He is a world known expert on this topic and has been invited to participate in programs and conferences both in Brazil as well as internationally.

From April 2 to September 28, 2012, Judge De Sanctis was a fellow at Federal Judicial Center in Washington, DC.

Since 2013, Judge De Sanctis has also been an Advisory Council member of the American University College of Law on its Program for Judicial and Legal Studies Brazil-United States.

His publications include, among others:

Money Laundering Through Art: A Criminal Justice Perspective. Heidelberg: Springer International Publishing Switzerland, 2013

"Recent Legal and Judicial Reform Initiatives in Brazil." International Judicial Monitor. Published by the International Judicial Academy, Washington, DC, with assistance from the American Society of International Law, Winter 2013, available at http://www.judicialmonitor.org/current/judicialreformreport.html

"Popular Action: Using Habeas Corpus in the Context of Financial Crimes" in Popular Action ("Ação popular: A utilização do habeas corpus na dinâmica dos crimes financeiros" in Ação Popular. São Paulo: Saraiva, 2013)

"Coherent and Functional Criminal Law" ("Direito Penal Coerente e Funcional" in *Revista dos Tribunais. Edição especial dos 100 anos.* Vol. 919. São Paulo: Revista dos Tribunais, May 2012)

"Telephone Tapping and Fundamental Rights," in *A Tribute to Afrânio Silva Jardim* ("Interceptações Telefônicas e Direitos Fundamentais" in *Tributo a Afrânio Silva Jardim: escritos e estudos.* Rio de Janeiro: Lúmen Júris, 2011)

Money Laundering through Gambling and Football. Analysis and Proposals (*Lavagem de* Dinheiro. Jogos de Azar e Futebol. Análise e proposições. Curitiba: Editora Juruá, 2010)

Criminal Liability of Corporations and Modern Criminal Methods (*Responsabilidade Penal das* Corporações e Criminalidade Moderna. São Paulo: Saraiva, 2009)

Organized Crime and the Disposal of Seized Assets: Money Laundering, Plea Bargains, and Social Responsability (Crime Organizado e Destinação de Bens Apreendidos. Lavagem de Dinheiro, Delação Premiada e Responsabilidade Social. São Paulo: Saraiva, 2009)

"The Constitution and Freedoms" in *Constitutional Limitations on Investigations* ("Constituição e Regime das Liberdades" in *Limites Constitucionais da Investigação.* Rogério Sanches Cunha, Pedro Taques and Luiz Flávio Gomes. São Paulo: Revista dos Tribunais, 2009)

The Fight Against Money Laundering: Theory and Practice (*Combate à Lavagem de Dinheiro, Teoria e Prática.* Campinas: Millennium, 2008)

"Human Trafficking: The Crime and Victim Consent" in *Women and Criminal Law* ("Tráfico Internacional de Pessoas: Tipo Penal e o Consentimento do Ofendido" in *Mulher e Dereito Penal.* A collaborative work. Rio de Janeiro: Forense, 2007)

"Crimes Against the National Financial System: A Precursor to Money Laundering" in *Money Laundering: Commentary on the Law by Judges at Specialized Courts* ("Crimes contra o Sistema Financerio Nacional como antecedentes de Lavagem de Valores" in *Lavagem de Dinheiro—Comentários à Lei pelos Juízes das Varas Especializadas.* A collaboratve work. Porto Alegre: Livraria do Advogado, 2007)

Criminal Tax Law: Highlights (*Direito Penal Tributário: Aspectos Relevantes.* Campinas: Bookseller, 2006)

Criminality in the National Financial System: Criminal Law and Protection of Brazil's National Financial System (*Punibilidade no Sistema Financeiro Nacional: Tipos Penais que Tutelam o Sistema Financeiro Nacional.* Campinas: Millennium, 2003)

Criminal Liability of Corporations: An Overview (*Responsabilidade Penal da Pessoa Jurídica,* São Paulo: Saraiva, 1999)

Judge De Sanctis has also written a number of articles published in newspapers and magazines specializing in law and economics.

Chapter 1
Introduction

Since ancient times, man has glorified athletes for their importance and for the beauty of their practice. The vast commercialization of sport in modern times, with its internationalized labor market and huge investments by media and sponsors, would have been unthinkable in previous eras. The existence of large sums of money with wealthy private investors who cross borders to fund sports would have been unimaginable.

The romantic view of sport has dissipated and no longer finds support or justification in today's world. Crime that was typically confined to certain sectors found its way into sport, transforming itself in order to carry out offenses that are extremely harmful to society.

Recognizing the opportunity for huge profits, money gradually discovered the world of sport and began to control it. On one hand, the increase in cash flow has allowed large numbers of people to access the world of sport through various investments. On the other hand, it has led to the harmful effects of fraud, tax evasion, corruption, doping, human trafficking, illegal gambling, match fixing, and money laundering. There is no doubt, therefore, about sport's vulnerability to a number of global threats.

It was not by accident that money laundering took such an unusual turn. Controls enacted pursuant to the recommendations by the Financial Action Task Force (FATF), aimed at cracking down on money laundering, made it necessary for criminals to seek out new mechanisms for the laundering of ill-gotten gains. Furthermore, the globalization of financial markets and the rapid development of information technology have gradually steered the underworld economy toward new possibilities for the commission of financial crimes.

Like so many other businesses, sport and gambling have been used by criminals to launder money and derive illegal income. As in the art world, criminals in the sport world are not always motivated by monetary gain. Social prestige, rubbing elbows with celebrities, and the prospect of dealing with authority figures may also attract private investors bent on skirting the law. Its high degree of specialization—inasmuch as few are really familiar with this market—could also contribute toward attracting illegal activities.

F. M. De Sanctis, *Football, Gambling, and Money Laundering*,
DOI 10.1007/978-3-319-05609-8_1, © Springer International Publishing Switzerland 2014

The absence of adequate and well-designed legislation gives power and mobility to organized crime, allowing its continuity and illegal acquisition of unprecedented amounts of wealth. Unreasonable, unjustified, and repeated tolerance by authorities toward criminal activities "practiced in the name of sport" has undermined the credibility of the sport industry. The inertia and inefficiency that plague enforcement in this industry must be dealt with through an assessment of sport regulation. Taking isolated and uncoordinated positions is irrational and runs serious risks. It is now more than ever necessary to use legal tools to bring an end to organized crime.

Because it is one of the most popular and celebrated sports in the world, and is also the subject of much attention and concern from many authorities, football will receive special attention and consideration in this book. Football is played by more than 265 million people in the world. According to the Fédération Internationale de Football Association (FIFA), there are 38 million professional players, duly registered, and about 301,000 clubs. Football has experienced an extraordinary growth since the early 1990s, a result of increased television rights and sponsorships. The market for professional players has experienced an unprecedented internationalization, allowing more and more transfers of resources at a transcontinental level.

The high volume of resources crossing boundaries and the lack of transparency in these transactions should demand a more incisive control by authorities, whose absence or ineffectiveness provides a unique opportunity for criminals to launder money. Yet, there is a true and apparent conflict. On one hand, there is a desire for autonomy in the organization and operation of sport. In Brazil, this autonomy is subject to a requirement of exhaustion in disciplinary and competition matters by a government judicial body called Sport Justice (Justiça Desportiva). On the other hand, there is an imperative need for state action in the prosecution of offenses committed by criminal organizations that rely on sport to perpetuate their actions and gain profit.

In football, image contracts, advertising contracts, and sponsorship contracts can be tools for criminal practice, notably tax evasion, since the money stipulated in these contracts is commonly transferred to accounts of companies in tax havens. This results in serious risks of fraud, since it is tempting to not declare the money received, which requires the use of third parties in various financial transactions. The most common form of cash payments involves jurisdictions located abroad, which allow the final destination of payments to be disguised. Image rights are also used to conceal the amounts actually paid to players.

Gambling, like sport, has a recreational role in society. Sport and gambling are similar in the sense that both are vulnerable to money laundering due to the lack of transparency and the attraction of large sums of money that characterize the two sectors. In addition, gambling is directly related to sport through betting on games and matches.

Therefore, lotteries, casinos, and gambling houses will also receive special attention in this book. The economic impact of the gambling sector is evident because large investments are channeled through it. There are also societal impacts, including business development and an extensive transmission of cultural values. Yet, the growth of this industry has encountered illegal practices, especially corruption, tax evasion, and money laundering.

The author's purpose in this book is to go beyond a mere introduction to this captivating topic. Considerations will be presented in an effort to further the study of methods likely to add transparency to business dealings and thereby inhibit or curtail unlawful activities. This book seeks to dispel the many mysteries surrounding certain practices. It will provide an understanding of the deficiencies in the world of football, as well as the weaknesses present in lotteries, casinos, and gambling houses. The aim is therefore to shed light on commercial practices in the world of sport in order to improve the system of crime prevention and punishment.

Some statutes and legislation deserve special analysis, as does the role of oversight bodies and regulators (FIFA, confederations or federations, clubs, and government agencies). The book will also identify legislative and institutional loopholes that might give power and mobility to organized crime, thereby making it a more deeply entrenched source of unprecedented illicit wealth.

The issue of violence in sport will also be addressed, for the crimes committed in the sport world are not limited to economic and financial ones. Sport provokes strong feelings in people that sometimes result in rash action. The emotional aspects of sport can, in a way, explain why people sometimes do not view it seriously in terms of its management and funding. In other words, authorities do not prioritize investigating financial crime in sport.

Betting on games has developed a sort of sophistication, with numerous operators working in several countries and using the Internet. This has increased the risk of illegal money laundering. Therefore, countries must regulate the gaming market so as to make it transparent because profiteers use countries that do not regulate or supervise games. It is not easy to control speculators who use online services and work from abroad. This, combined with the lack of transparency in the market, makes it an ever more attractive vehicle for criminals.

It is important to note that FIFA, the world governing body of football, which also has the mission of regulation, promotion, and development throughout the world, has made efforts to prevent unlawful practices in football. National associations, federations, and confederations are also properly engaged in these efforts. They should do more to offer necessary support, like professional training on money laundering, so as to enable communication on suspicious transactions. They should also analyze regulations and disciplinary actions of those in charge of managing and protecting sport.

The FATF, an intergovernmental body linked to the Organization for Economic Cooperation and Development, has shown great concern in the prevention of money laundering in sport.

For a global vision of the role of each actor in the football industry, it is important to study, among others, the Brazilian bodies of discipline and control, including the Brazilian Football Confederation, regional federations, clubs, the Council for Financial Activities Control (Brazil's financial intelligence unit), the Federal Revenue of Brazil, police, public prosecutors, and the Sport Ministry. They could all, in theory, coordinate actions to prevent crime.

After connecting the dots that make up this interesting topic, the book suggests effective proposals for improving the assessment, investigation, and prosecution of the crime of money laundering. The book takes into account both the legislation and the immediate needs of the bodies involved, judicial or otherwise. The world is in a pivotal moment. Brazil, one of the greatest powers in football, is especially in a critical time, as it will soon host both the World Cup and the Olympics.

Gambling and sport have demanded staunch fair play to prohibit common violence. But it is also essential that they develop "financial fair play" to curb economic and financial crimes that can, in the not so distant future, compromise and bring down the activity itself. This book can assist global authorities in taking necessary preventive measures.

We must be mindful that one of the essential criminological features inherent in money laundering, as Pedro Caeiro, citing Jorge Fernandes Godinho and Luís Goes Pinheiro, reminds us, is its necessary links to organized crime, which in turn add considerable diversity to the types of conduct that its prosecution and enforcement may prevent.[1]

Our aim is to provide a reading on this sector, a snapshot of the market, which will provide the groundwork and guidance necessary to give it transparency and a backdrop sufficient for a particularized analysis. Some rigor in procedures for cataloging and investigation are in order, for we ought to remember that the resurgence of organized crime is often the result of a systemic atmosphere of inattention, mutual tolerance, and ethical codes that, however lofty, are in practice applied only selectively. Matters are worsened by the arrogance and permissiveness, if not covert complicity, of portions of civil society (the elite, the press, etc.) that insist on pointing out only the defects that do not suit their purposes.

This book is divided into nine chapters. Chapter 2 addresses illegal activities in the sport world. Chapter 3 deals specifically with financial crime in football. It also provides an overview of the French experience of controlling accounts in football. Chapter 4 discusses overarching topics of gambling and lotteries, measures for crime prevention, and enforcement agencies. Chapter 5 is about illegal betting. Here, important cases that were recently covered by the media will be discussed. Chapter 6 seeks to identify the use of illegal and disguised instruments for payments in business transactions. Forms of payment and the use of nongovernmental and offshore organizations are addressed. International legal cooperation and asset forfeiture are analyzed in Chapter 7. Chapter 8 covers conclusions that may go a long way toward clarifying how the prevention of money laundering applies to the gambling and sport industries. The final chapter, Chapter 9, covers national and international proposals for improving the industry so as to prevent money laundering and the financing of terrorism.

---

[1] Cf. Pedro Caeiro, in *Branqueamento de capitais*. Manual distributed in a course sponsored by the Organization of American States (OAS) and the Brazilian Ministry of Justice and presented to the Brazilian judges and prosecutors on October 17–21, 2005, p. 4.

Bibliography

CAEIRO, Pedro, in *Branqueamento de capitais*. Manual distributed in a course sponsored by the OAS and the Brazilian Ministry of Justice and presented to Brazilian judges and prosecutors on October 17–21, 2005, p. 4.

Chapter 2
Sport

2.1 Sport and Cultural Heritage

The world of sport is a paradox. On the one hand, the organization and functioning of sport runs on the principle of autonomy, such that leaders of sports organizations or relevant associations require competition and discipline matters to go through specialized sport courts, where they exist (like Brazil), before involving local authorities. On the other hand, it is imperative to prosecute offenses committed by criminal organizations that use the sport sector to cover up and facilitate their actions, and this requires the involvement of local authorities.

Indeed, the sport industry is one of many sectors that has attracted perpetuators of financial crimes, such as money laundering. The traditional, simplistic view of the sport sector as an area where no actions are taken in bad faith by those involved is no longer accurate. As several criminal cases are reported and more are gradually revealed to the public, the traditional view should cease to exist.

Yet, some people continue to have this innocent perception, where they refuse to acknowledge organized crime in sport and show a lack of interest in assessing the resources within the sport sector that are used for organized crime. This innocent perception threatens the sport sector and will even compromise the state order. This romantic view of sport does not have support in today's world, where crime that was once confined to a few sectors has expanded to the sport arena. Moreover, this expansion has transformed the sport sector's operations to propagate crimes of extreme social harmfulness.

Since ancient times, man has given importance to sport and idolized its athletes. This esteemed view of sport has led to worldwide marketing of different games, an unprecedented international labor market surrounding sport, and large sums of money invested by the media, various sponsors, and wealthy private investors. One can certainly say that money gradually began to discover sport and control it. While this increase in cash flow allows a great amount of people to contribute diverse investments, it can also lead to risks of fraud, tax evasion, and corruption. The movement of money within the sport sector can also become a channel for laundering dirty money.

F. M. De Sanctis, *Football, Gambling, and Money Laundering*,
DOI 10.1007/978-3-319-05609-8_2, © Springer International Publishing Switzerland 2014

There is no doubt that the sport sector is vulnerable to a number of issues and global threats. Over time, authorities have become aware of problems within the sports industry, which include racism, corruption, doping, human trafficking, illegal gambling, match fixing, violence, and money laundering. The high volume of resources being transported across borders and, as we will be discussed later, the lack of transparency in such negotiations require greater control by authorities. The absence of such control provides unprecedented opportunities for organized crime to launder dirty money within the sport sector. All of this leaves out the hunger for profits of private investors, who view the sport sector as just another commercial business whose first principle would be "business is not built on the Beatitudes."

Despite the existence of international media reports and various studies pointing out the international schemes of organized crime within the sport sector, many choose to ignore these reports. Perhaps this is done due to denial or as a matter of convenience by those managing or participating in the sport sector. This willful ignorance prevents people within the sport sector from seeing any connection between organized crime and sports, and such denial only stimulates more crime due to state inaction.

There seems to be a general perplexity surrounding crime in sport that has cooled efforts of state control, which is dependent on confidential information provided by clubs or sports associations that is not always easily understood. This realization has surprised scholars and confronted the orthodox practices of sports management. The sport industry is surrounded by an aura of mystery that cannot be boiled down to a simple prognosis of unlawful behavior. Instead, it will take a differentiated approach to uncover the criminal acts in the sport sector.

2.2 Structural Violence and Organized Crime

Notwithstanding the central focus of this study—financial crime in the sport sector—violent crime is a constant threat to individuals at sporting events, as well as to their assets. For example, homicides, thefts, and robberies are still the crimes that most people fear while at sporting events. Street robberies make up a large portion of these criminal offenses, threatening the general public as well as foreigners who attend international sporting events. Street robberies are particularly dangerous due to the frequent use of firearms to carry out such crimes, especially in underdeveloped countries.

Football has particularly engendered violent forms of aggression, both in stadiums and outside stadiums on the streets. For example, on November 9, 1947, on the Brazilian field at Teixeira de Castro, the team from Bonsucesso defeated the team from Fluminense by a score of four to three. The referee, Alberto da Gama Malcher, was brutally attacked by six fans who disagreed with his ruling. The attack resulted in several injuries.

The congregation of large groups of people at sporting events allows for the enjoyable and spectacular experience of watching a game together and the collective manifestations of various cultures. However, this congregation of fans can also lead to a "mob mentality," where people engage in negative attitudes or actions as a group that they would never do in isolation. The stirring up of emotions that leads to the united gathering and support of fans can also attract those who wish to act out negatively on their emotions in a way that is normally not allowed in other social contexts. Any negative emotions that are socially repressed may find an outlet at sporting events due to the safety of numbers, causing fans to act on their frustrations in a way that would not happen if they were alone. Moreover, the stadiums in some countries are badly maintained and difficult to exit, thus creating conducive environments for festering annoyances to be aggravated until they erupt into episodes of physical violence.

Homeland security is a major concern for the general public, and issues of security are especially salient when it comes to football. On game days, authorities make plans that are meant to curb violence in football games, particularly in the stadiums, in surrounding areas, and sometimes in different places outside of game sites. However, fans and their families are skeptical of these measures due to the apparent inability of the state to reduce crime and provide greater safety at football games.

In most cases, carrying out policies that are meant to contain such crimes proves to be a challenge to government agencies. Such offenses are still considered a major threat to the citizens, which shows that what many countries are doing to contain such crime is not enough. However, the prevalence of such violent crime is based on a combination of factors that depend on the socioeconomic conditions surrounding the sport environment, and any deficiencies in police work should not be regarded as the sole cause of such crimes.

People and players are also fully aware of the serious problems regarding crime within football and the risks that stem from the absence, or insufficient action, of local authorities. The players themselves have also been victims of violence.[1] The growth of the underworld within the football arena has caused gangsters to become overconfident.

Furthermore, other crimes have been committed within the scope of international sporting events, such as Internet crimes, drugs, human trafficking for sexual exploitation of women and children, and terrorism. Misguided public policies have moved individuals and supporters away from direct participation in the management of sports, especially with respect to the revenues made by sporting clubs. These policies have made it harder to stop crime within the sport sector, since less people are able to infiltrate and prevent such crime.

On January 8, 2010, the world was surprised by the attack on the team bus of the Togo national football team, which was carried out by the Liberation Front of Cabinda and led to the killing of three people and the injuring of several others. This

---

[1] Graham Johnson reveals that "in the past, good players expected fans to get upset when they left a club, especially if they went to a big rival. But today, the boos and hisses of the ordinary supporters are accompanied by a sinister undertone: death threats, hate campaigns and attacks on family members." In Football and Gangsters: How Organized Crime Controls the Beautiful Game. Great Britain: Cox and Wyman Ltd., 2007, p. 61.

tragic event demonstrates how sport can be used as a target for terrorist purposes. Such attacks are not small or isolated acts carried out by unknown political rebels. Rather, they represent a global threat to the sport sector by known terrorist groups who want to use football as a platform to spread their message. This fear goes beyond the fear associated with normal levels of violence within sporting events.

According to Graham Johnson, "football and gangsters is still an unspeakable taboo. Gangsters often use sophisticated methods of entrapment, including drugs, prostitutes and 'honeytraps' to blackmail players, who are often too young, too scared or too stupid to be wary in the first place or to report incidents of extortion to the police afterwards."[2]

Thus, the sport sector has become an area ripe for various criminal activities, with a higher incidence of common crimes and a potential threat of more serious terrorist activities. It should be noted that Brazil has not been directly threatened by terrorist practices. However, this fact does not justify forgoing constant vigilance in attempting to prevent such practices, particularly when hosting international sporting events.

One way to reduce such crime is to create task forces that plan, coordinate, and involve cooperation among various actors, including educators and nongovernmental organizations (NGOs), to respond to such crime. These task forces are presented with a very difficult and complicated task; therefore, their policies should not solely involve the increased use of police forces. Even if they can plan specific safety measures for world sporting events, the apparent inability of local authorities to deal with common crime suggests that people from other lines of work need to be involved to address the problem. That being said, the inability to eradicate common crime by local authorities does not mean that there has been no success in driving public policy to deter wrongdoing. This can be seen in the achievements in public safety by Brazil during the Pan American Olympic Games held in Rio de Janeiro in 2007.

Unlike international sporting events, where there is greater knowledge on what needs to be addressed in order to create a successful policy, it is hard to control the socioeconomic conditions that typically cause general crime. In addition to violence in sports, the will to win has led some athletes to engage in another kind of criminal activity, doping—the use of substances that improve physical performance. This is not a new concept, as the practice of doping is even older than the Olympics. Four thousand years ago, the Chinese, well aware of the effects of the tea plant called "machuang," which contains ephedrine in high doses, used this plant to increase their working capacity. During the Olympic Games of Antiquity in 800 BC, athletes drank various herbal teas and oils and used mushrooms to enhance performance. In the nineteenth century, it became popular among athletes to consume a drink called "Vin Mariani," which was made from the leaves of cocaine and took the name of the alchemist who produced it.

The modern Olympics, an event of both social and national celebration, emphasizes victory so much that it has created a culture of obtaining wins at any price.

---

[2] In Football and Gangsters. How Organized Crime Controls the Beautiful Game. Great Britain: Cox and Wyman Ltd., 2007, p. 7.

As Luis Fernando Correia points out, "the first games of the modern era, organized by Baron Coubertin in Athens, in 1896, marked the emergence of 'small balls', spheres containing various stimulants as cocaine, ephedrine and strychnine."[3] This emphasis on winning led athletes to their deaths through excessive use of stimulants and hormones during the 1960 and 1964 games. Despite the International Olympic Committee's strong attempts to control the use of doping, frequent cases are still recorded. For example, seven cases of doping were reported in the 2004 Athens Paralympics.

As it has grown in recent years, the sport sector has reflected an exacerbation of passion and expression, especially in football. Assaults are a constant in the sport, but it has also drawn attention to cases of racism[4] and violence by organized supporters.[5] According to Graham Johnson, "in countries where armed gangs still rule the roost, they have used football to vent their violence."

A paradigmatic case of racism occurred in Brazil in April 2005, when an Argentine player, Leandro Desábato of Quilmes, made racist comments toward Grafite, a soccer player of São Paulo, at the Morumbi stadium. The altercation that arose between these two players led to the entry of a police chief into the field to arrest the Argentine player. In another incident, soccer player Roberto Carlos was taunted by fans of Atletico Madrid who called him "monkey." This resulted in a fine of US$ 787 to the club, which had already been fined by the European Union of Football at US$ 12,000 after fans of the club also made racist gestures toward Brazilian defenders Juan and Roque Junior. In Italy, striker Mario Balotelli from Inter Milan suffered racial discrimination by Juventus fans in 2009.

The situation regarding racism in Europe has reached such high levels that an association named "Red Card" was created in Barcelona, Spain, to combat racism by mobilizing clubs and athletes against this phenomenon and taking legal action to ensure respect for minorities in sports.

The Fédération Internationale de Football Association (FIFA) announced on May 6, 2013, that it would also take measures to combat racism in football. For instance, one FIFA employee will be present during each game to help the referee handle racist encounters. In 2006, the Brazilian Football Confederation (CBF) published resolutions that punish clubs that practice or tolerate racism among their staff or fans. Football clubs can now be punished with financial or sport sanctions for engaging in racism.[6]

---

[3] *In*: A história do doping nos esportes. Available at http://g1.globo.com/Noticias/Ciencia/0,,MUL1267929-5603,00-A+HISTORIA+DO+DOPING+NOS+ESPORTES.html.

[4] According to Steve Greenfield and Guy Osborn, "The football authorities are very conscious of the problem of racism within the game, although there have been highly public instances of alleged abuse with racial undertones between players that have been largely ignored." In The Football (Offences and Disorder) Act 1999: Amending s3 of the Football Offences Act 1991. 5 J. C. L. 55 2000, p. 62.

[5] Football and Gangsters. How organized crime controls the beautiful game. Great Britain: Cox and Wyman Ltd., 2007, p. 195.

[6] In Racismo poderá gerar até rebaixamento. FIFA. Federação cria pacote de propostas para combater discriminação, com punições financeiras e esportivas. Folha de São Paulo, esporte, D3, May 7, 2013.

The root of racism in football may stem from the fact that only white people and aristocrats played it in the beginning. Considered to be the sport of the Brazilian elite in the 1920s, the makeup of football players changed in 1923, when a new football team, Vasco da Gama, was created to destabilize the hegemony of the ruling classes. The team, coming from Rio de Janeiro, was composed of black players, multiracial players ("mulattoes"), and white players from lower-income origins, and it won the State football championship during its first year.

In addition to its privileged origins, there is also the belief that it is appropriate to release any latent aggression without limitations while engaged in sports. This view makes sporting arenas analogous to lawless lands, where human aggression finds legitimacy to express itself. While "free expression" is a fundamental value, it must yield to the right to live in a peaceful society in which human dignity is always protected. Free expression cannot undermine the foundations of democracy and open ground for intolerance and totalitarianism.

The release of the basest emotions during sporting events has not only led to assaults and racist encounters but also been leveraged by people who "organize" militarily, causing the spread of new kinds of physical violence in cities. These instigators of "organized" violence are able to lure in younger fans in particular. The possibility of escaping from reality, combined with a manifestation of power that includes congregating with others and retaining anonymity, influence younger fans to perform every kind of expression of intolerance and terror under the illusory cloak of supporting their favorite team. Theories that explain this adolescent rebellion include the desire to reject parental domination, sport being a release for all kinds of adolescent frustrations, and an antiestablishment belief in opposing the domination of the government or ruling party.

The possibility of strict liability for the leaders of organized violence based on the actions taken by its members may eventually constitute effective means to steer these leaders and their members toward lawful action. It is not fair that leaders have access to privileges, like tickets, participation in meetings, transportation, etc., without also being liable for the violence they cause.

In its most trivial sense, there are currently several cases of vandalism, foolishness, and irrational behavior, to the point that organized groups of fans are only allowed to move around with police escorts. On the other hand, there are also more serious instances of organized violence, where groups are usually armed with iron bars, incendiary instruments, and traditional weapons, and war cries can be heard. The shouts of these groups represent intolerance and hostility at its worst, where the leaders opine ideological appeals ostensibly based on reason instead of acknowledging their true racist and prejudiced purposes. The brutal scenes in the episode at the Pacaembu stadium in São Paulo in 1995, where supporters of teams from São Paulo and Palmeiras clubs staged regrettable confrontations that were televised, showcase the intensity of serious organized violence.

Even in more developed countries, this "hooliganism," which originally stemmed from the working classes and subcultural groups, has expanded to the general population. Even the groups in more developed countries express their worldview about the differences among various groups in a negative light through slogans, gestures, and songs. In England and Spain, the issue of organized violence among "hooligans" has been a matter of public safety since the 1980s.

Football mobilizes the masses and is a trendsetting sport; however, it also models aggressive behaviors and provides an environment to act out on these aggressions. Even leaders of various sports teams and organizations may reflect and support the implicit understanding that anything is acceptable in the sports environment. One cannot forget, for example, the incident where the former coach of the Brazilian national team, Carlos Alberto Parreira, spit water served by Argentine players in an obvious display of incivility. This "everything is acceptable" attitude can often facilitate crime and calculated disorder, since such an attitude makes little distinction between legal and illegal actions, especially in the economic field. However, these same sports leaders should recognize that the danger of violence and crime that is constantly present during these football games will eventually affect the sport by compromising its popularity. National policies to combat violence in sport, including organized violence, have been introduced. In Spain, such policies began to surface in 1987, while similar policies cropped up in England in 1990. In Brazil, the Charter of Brasilia was drafted in 2003 after debates sponsored by the Sports Ministry took place among people involved with football. In 2004, a national commission was created that dealt with violence in sports and required ongoing action to address this problem. Unfortunately, despite these efforts, the main values that are meant to be instilled in sport—that is, coexistence, brotherhood, and welfare—remain shattered and weakened by attitudes that reaffirm cultural, social, and economic differences and attempt to give certain groups privileges and advantages over others.

2.3   Unlawful Practice and Consent

The private appropriation of the sport industry can be observed by the actions taken by the ex-president of the CBF, as evidenced by the frequent comments and assertions made by his top managers close to him. The ex-president threatened to strip a group of Brazilian teams called "Club 13" of television contract rights if his nominee for election was not chosen to command this entity. These competitors claimed that the dispute affected the very survival of these sports clubs.[7] Common sense goes against taking such a stance, trumpeted so shamelessly, revealing the sad state of the football industry.

Besides the activities of bookies and middlemen outside the sport who seek illicit profits at the expense of the pain and suffering of athletes and their families, there

---

[7] Cf. ARRUDA, Eduardo. *A ameaça. Folha de São Paulo*, Esporte, D2, Painel FC, March 31, 2010.

is a structured problem involving the industry of football as a whole.[8] For example, countries prefer to use service contracts in professional football clubs based on the desire to not contribute to the players' pensions and to not make social security payments. According to Robert Siekmann, "the player is regarded as an artist. Football is thus treated as a special branch of industry that simply cannot be compared with ordinary business."[9]

The US Congress passed the Sports Agent Responsibility and Trust Act ("SPARTA")[10] in 2004. It is the only federal legislation specifically dealing with sports agents. According to Timothy G. Nelson, SPARTA "was enacted to protect colleges and universities and, to a lesser extent, student-athletes from improper recruitment practices of agents. Among other things, the law makes it illegal for an athlete agent to provide anything of value to a student athlete or anyone associated with the student athlete before the student athlete enters into an agency contract, including any consideration in the form of a loan, or acting in the capacity of a guarantor or co-guarantor for any debt."[11]

Often, licensed agents use web sites and unauthorized persons for the practice of "dirty" service. Even big clubs make use of smaller clubs, usually located in the countryside, for the placement of children who have not yet reached the legal age to practice sports. Common reports show the prevalence of sports leaders breaking laws and regulations established by FIFA in order to obtain a player with a promising future. There are even cases of people linked to human or drug trafficking who traffic young football players for large amounts of money.

There are some loopholes to the rules that prohibit the use of minors in international transfers in football. For example, Arts. 19 and 19bis in the FIFA Regulation on the Status and Transfer of Players, of October 19, 2003, called "Protection of minors,"[12] actually facilitate criminal activity because they represent a relaxation of the rules in a way that makes them easy to circumvent.

Title VI, titled "International Transfers Involving Minors," states the following:

---

[8] In an interesting case, *Eastham v. Newcastle United Football Club*, George Eastham, a famous British football player, was at the end of his contract with Newcastle United. He requested to be transferred to another club, but at that time a club could effectively hold a player for life by withholding his playing registration and refusing any and all transfer requests by both the player and other clubs. In this system, considered an "archaic one," players were like "soccer slaves." In: James G. Irving, *Red Card: The Battle Over European Football's Transfer System*. 56 U. Miami L. Ver. 667, 2002, p. 670.

[9] In Labour Law, the Provision of Services, Transfer Rights and Social Dialogue in Professional Football in Europe. 4 ESLJ 1 2006–2007.

[10] 15 U.S.C. §§ 7801–7807 (2006).

[11] In Flag on the Play: The Ineffectiveness of Athlete-Agent Laws and Regulations—and How North Carolina Can Take Advantage of a Scandal To Be A Model For Reform. 90 N.C. L. Rev. 800, 2011–2012, p. 803.

[12] FIFA Regulations on the Status and Transfer of Players, http://www.fifa.com/mm/document/affederation/administration/01/95/83/85//regulationsstatusandtransfer_e.pdf, accessed on June 21, 2013.

Art. 19 Protection of minors

1. International transfers of players are only permitted if the player is more than the age of 18 years.
2. The following three exceptions to this rule apply:
    (a) The player's parents move to the country in which the new club is located for reasons not linked to football.
    (b) The transfer takes place within the territory of the EU or European Economic Area (EEA) and the player is aged between 16 and 18 years. In this case, the new club must fulfil the following minimum obligations:
        (i)   It shall provide the player with an adequate football education and/or training in line with the highest national standards.
        (ii)   It shall guarantee the player an academic and/or school and/or vocational education and/or training, in addition to his football education and/or training, which will allow the player to pursue a career other than football, should he cease playing professional football.
        (iii) It shall make all arrangements necessary to ensure that the player is looked after in the best possible way (optimum living standards with a host family or in club accommodation, appointment of a mentor at the club, etc.).
        (iv)   It shall, on registration of such a player, provide the relevant association with proof that it is complying with the aforementioned obligations.
    (c) The player lives no further than 50 km from a national border and the club with which the player wishes to be registered in the neighboring association is also within 50 km of that border. The maximum distance between the player's domicile and the club's headquarters shall be 100 km. In such cases, the player must continue to live at home and the two associations for their concern must give explicit consent.
3. The conditions of this article also shall apply to any player who has never previously been registered with a club and is not a national of the country in which he wishes to be registered for the first time.
4. Every international transfer according to paragraph 2 and every first registration according to paragraph 3 is subject to the approval of the subcommittee appointed by the Players' Status Committee for that purpose. The application for approval shall be submitted by the association that wishes to register the player. The former association shall be given the opportunity to submit its position. The subcommittee's approval shall be obtained prior to any request from an association for an International Transfer Certificate and/or a first registration. Any violations of this provision will be sanctioned by the Disciplinary Committee in Accordance with the FIFA Disciplinary Code. In addition to the association that failed to apply to the subcommittee, sanctions may also be imposed on the former association for issuing an International Transfer Certificate without the approval of the subcommittee, as well as on the clubs that reached an agreement for the transfer of a minor.

19bis Registration and reporting of minors at academies:

1. Clubs that operate an academy with legal, financial, or de facto links to the club are obliged to report all minors who attend the academy to the association upon whose territory the academy operates.
2. Each association is obliged to ensure that all academies without legal, financial, or de facto links to a club:

    (a) Run a club that participates in the relevant national championships; all players shall be reported to the association upon whose territory the academy operates, or registered with the club itself.

    (b) Report all minors who attend the academy for the purpose of training to the association upon whose territory the academy operates.

3. Each association shall keep a register comprising the names and dates of birth of the minors who have been reported to it by the clubs or academies.

4. Through the act of reporting, academies and players undertake to practice football in accordance with the FIFA Statutes, and to respect and promote the ethical principles of organized football.

5. Any violations of this provision will be sanctioned by the Disciplinary Committee in accordance with the FIFA Disciplinary Code.

6. Art. 19 shall also apply to the reporting of all minor players who are not nationals of the country in which they wish to be reported.

The text says nothing about transfers within a country, thus failing to preclude the spurious use of these forms by organized crime.

Furthermore, two of the three exceptions to the general rule prohibiting international transfers include situations where one can easily violate the standard rules.

The first, Art. 19(2)(a), permits international transfer when the parents of a young player move to the country in which the new club is located for reasons not linked to football. It is possible for clubs to use this exception and facilitate a parent's move in order to remove any obstacles to obtaining the young player. Thus, there are situations where clubs or agents have admitted to employing a young player's father as the stadium gardener or, in another case, as a driver for the club in order to circumvent the general rule and take advantage of the talented young player. A parent moving for a new job allows the club to hire the parent's younger child, even if the move is, in fact, linked to football and was solely done to allow a teen to get player status.

The other exception, Art. 19(2)(b), permits transfer within the EU or the EEA if the player is between the ages of 16 and 18 years, as long as the player is placed in optimal living standards and provided adequate football training and academic or vocational education. However, there are reports of immigration policies that allow a 16-year-old child to come to Europe, only to be transferred by a football club, thus triggering this exception to the general rule.

The FIFA regulations also require clubs that run football training schools to report the names of all the minors participating in such clubs to the appropriate association or local federation (Art. 19bis(1)). Even when there is no legal or de facto financial link between the training schools and the clubs, these schools are still required to report to the mentioned entities all listed minors within the school (Art. 19bis(2)). Moreover, the failure to record the names and dates of birth of all minors within these football training schools results in sanctions against the violating entities (Art. 19bis(3)).

Even though registrations and international transfers of players are subject to the approval of the subcommittee appointed by the Players' Status Committee (Art. 19(4)), this provision and its accompanying sanctions for those who violate it have not curbed the abuses that are committed and often reported in this area of sport. Despite these ostensibly stringent rules, clubs often use their lawyers to find loopholes that would allow them to permanently overcome any legal barriers, even though these barriers exist to protect adolescent players. In addition to the above exceptions being manipulated by clubs and embassy officials, there have also been instances of passports being falsely created or modified to say that the players are older than they actually are.

These football training schools can be a secure mechanism to find a talent at an early age, with some schools being established in Africa, Asia, and Europe to enable the formation and movement of young players, even when such movement entails removing the player from his city and family. Children as young as 8 years old leave their home countries in Africa and Asia to train for European football, but in the end, only the best are employed by a football club in that continent. It is estimated that 99 out of every 100 applicants fail to climb their way to the big European clubs.

Furthermore, some players' careers are ruined by their agents, partying, or even the clubs themselves. One example of this is the story of a Senegalese player. Considered to be one of the most talented football players in his country, this 21-year-old young man left Senegal in 2005, hoping to play in European football. However, an agent sent the player to Norway for a trainee program on a relatively unknown team. After training for only a week, the club decided to send him back to Africa because he was not deemed good enough. He asked the club to send him to Italy, but he later disappeared. Three years later, he was found on the streets of Belgium.

The spirit of sport has surrendered to the thirst for financial gains, which has led to sports officials channeling resources to and from organized crime. Given the high salaries that are paid to players, clubs have sought to operate with cheap labor, which has produced a range of young people who are left unemployed and without a place to call home. In addition, foreign capitalists invested large sums of money in European clubs, thus inflating the value of market players. These capitalists are left not knowing the true amount of their investment or how much return they should have on their investment.

João Havelange, the Brazilian who was the honorary president of FIFA and who led the football world for 24 years, resigned on April 18, 2013, 12 days before being reported for bribery charges. According to Leandro Colon and Marcel Rizzo, he resigned in order to avoid the embarrassment of being punished by FIFA. On April 28, 2013, the FIFA Ethics Committee, created in July 2012, released the final report in the case. It revealed that Havelange, Ricardo Teixeira (the former president of the CBF), and Nicolás Leoz (Paraguayan ex-President of the Confederación Sudamericana de Fútbol (CONMEBOL)) were taking bribes from International Sport and Leisure (ISL), a Swiss sports marketing company. The bankrupt ISL was investigated by the Swiss Trial Court for paying bribes to officials in exchange for ease in obtaining contracts. In 2012, it was reported that Teixeira and Havelange received US$ 22.5 million in bribes.[13] In his conclusion, published on the FIFA web site, the judge of the Ethics Committee, Hans-Joachim Eckert, stated that the mentioned leaders received bribes from ISL between 1992 and 2000, but notes that,

---

[13] Cf. Havelange, 96, renuncia a cargo na Fifa para não sofrer punição. Folha de São Paulo, Esporte, D4, May 1, 2013.

at the time that these bribes took place, there was not a code of conduct within the entity.[14] It is noteworthy that the Code of Ethics only came into existence on June 10, 2004, and was later revised on September 15, 2006.[15]

While FIFA has made attempts to control financial crime in football, this oversight is still not enough to regulate the economic loopholes that make it fairly easy for new entrants in organized crime, with the help of their financial and legal advisors, to do whatever they want.

One important issue must be highlighted: the malpractice committed by professionals in the sport sector. Such practices often amount to criminal offenses, yet they are often committed without threat of punishment, or in certain cases, even with the consent of the alleged victims. A simple solution to this problem would be to prohibit from the sport industry those who violate their legal obligations in the exercise of their profession. This would extend to accountants, various public employees, and independent contractors. Brazilian law contemplates some possibilities of penalizing such professionals in Arts. 47 and 56 of its Penal Code.

During the sentencing portion of a criminal trial involving such crimes, the judge should consider, along with the reasons and circumstances behind the crime and the consequences of the crime, the victim's behavior to determine whether it is indicative of collusion with or acquiescence to the alleged perpetrators of the crime. The penalty should also be increased if a vile motivation exists, or if an abuse of power or breach of duty occurred during the undertaking of official or public duties.

The possibilities of criminal acts within the sport sector go beyond financial crime, as evidenced by the creation of fake documents, tax fraud, and instances of young players being kidnapped and even reduced to a condition analogous to slavery. Brazilian legislation enacted to protect children and adolescents highlights these offenses, which include the deprivation of liberty when children are delivered to third parties through payments or rewards, when parties are aided in the illegal transport of children abroad, and when these entities facilitate in the corruption of the minors themselves.[16]

On the other hand, human trafficking offenses are still a major focus of sports authorities due to their great international impact. International human trafficking is currently considered to be the third most profitable illicit activity, with drug trafficking and weapons trafficking constituting the two more profitable criminal ventures. Human trafficking now ranks second among transnational crimes, exceeding the prevalence of weapons smuggling, with drug trafficking still constituting the top transnational crime. It is estimated that about 700,000 women and 1,000,000 children are trafficked annually, where 92 % become victims of sexual exploitation and 21 % are forced into child labor. There are approximately 30 trafficking routes,

---

[14] http://pt.fifa.com/aboutfifa/organisation/president/news/newsid=2066172/index.html?intcmp=fifacom_hp_module_about_fifa, accessed May 1, 2013.

[15] In Code of Ethics, http://www.fifa.com/aboutfifa/organisation/footballgovernance/codeethics.html, accessed June 1, 2013.

[16] Statute of Children and Adolescents, Law No. 8.069, July 13, 1990, items 230, 238, 239, and 244-B.

with each victim generating a profit of US$ 50,000. This human trafficking leads to disastrous consequences—for every 100 trafficked humans, 24 contract a sexually transmitted disease, 3 contract human immunodeficiency virus (HIV), 15 women become pregnant, 26 suffer physical assaults, 19 suffer sexual assaults, and 9 are subjected to threats and intimidation.[17]

The following international documents on human trafficking show the history of attempts to prevent this long-existing crime:

1. **1904:** International Agreement for the Suppression of the White Slave Traffic
2. **1910:** International Convention for the Suppression of the White Slave Traffic
3. **1921:** International Convention for the Suppression of the Traffic in Women and Children
4. **1933:** International Convention for the Suppression of the Traffic in Women of Full Age
5. **1948:** Universal Declaration of Human Rights (Art. III: "Everyone has the right to life, liberty and security of person;" Art. IV: "No one shall be held in slavery or servitude; slavery and the slave trade shall be prohibited in all their forms")
6. **1949:** Convention and Final Protocol for the Suppression of the Traffic in Persons and of the Exploitation of the Prostitution of Others
7. **1966:** International Covenant on Civil and Political Rights (ratified by Brazil January 24, 1992; Art. 8: "1. No one shall be held in slavery; slavery and the slave-trade in all their forms shall be prohibited. 2. No one shall be held in servitude." Art. 9: "1. Everyone has the right to liberty and security of person.")
8. **1966:** International Covenant on Economic, Social and Cultural Rights (ratified by Brazil January 24, 1992; Art. 10: "The States Parties to the present Covenant recognize that 1. The widest possible protection and assistance should be accorded to the family, which is the natural and fundamental group unit of society, particularly for its establishment and while it is responsible for the care and education of dependent children.")
9. **1969:** American Convention on Human Rights, Pact of San Jose, Costa Rica (ratified by Brazil on September 25, 1992; Art. 6: "Freedom from slavery 1. No one shall be subject to slavery or to involuntary servitude, which are prohibited in all their forms, as are the slave trade and traffic in women.")
10. **1988:** Additional Protocol to the American Convention on Human Rights in the area of Economic, Social and Cultural Rights (ratified by Brazil on August 21, 1996; Art. 15: "Right to the Formation and the Protection of Families 1. The family is the natural and fundamental element of society and ought to be protected by the State, which should see to the improvement of its spiritual and material conditions.")
11. **1979:** Convention on the Elimination of All Forms of Discrimination against Women (ratified by Brazil on February 1, 1984)

---

[17] Data revealed by Fernando Capez, based on the 10th UN Sessions Commission for Preventing Crime and Criminal Justice, which took place May 13–22, 2003, in Vienna, Austria. The principal discussed theme was international women and children trafficking (*in: Curso de Direito Penal— parte especial*, p. 105).

12. **1989:** Convention on the Rights of the Child
13. **1994:** Inter-American Convention on International Traffic in Minors
14. **1994:** Inter-American Convention on the Prevention, Punishment and Eradication of Violence against Women, adopted by the Organization of American States (ratified by Brazil on November 27, 1995; Art. 1: "For the purposes of this Convention, violence against women shall be understood as any act or conduct, based on gender, which causes death or physical, sexual or psychological harm or suffering to women, whether in the public or the private sphere." Art. 2: "Violence against women shall be understood to include physical, sexual and psychological violence: a. that occurs within the family or domestic unit or within any other interpersonal relationship, whether or not the perpetrator shares or has shared the same residence with the woman, including, among others, rape, battery and sexual abuse; b. that occurs in the community and is perpetrated by any person, including, among others, rape, sexual abuse, torture, trafficking in persons, forced prostitution, kidnapping and sexual harassment in the workplace, as well as in educational institutions, health facilities or any other place; and c. that is perpetrated or condoned by the state or its agents regardless of where it occurs." Art. 3: "Every woman has the right to be free from violence in both the public and private spheres.")
15. **1998:** Rome Statute of the International Criminal Court (ICC; Art. 7: slavery, rape, enforced prostitution, or any other form of exploitation of considerable gravity, when "committed as part of a widespread or systematic attack against any civilian population, with knowledge of such an attack" is a crime against humanity)
16. **2000:** Protocol to Prevent, Suppress and Punish Trafficking in Persons, Especially Women and Children, supplementing the UN Convention against Transnational Organized Crime
17. **2000:** Optional Protocol to the Convention on the Rights of the Child on the sale of children, child prostitution, and child pornography

The Protocol to Prevent, Suppress and Punish Trafficking in Persons, Especially Women and Children (see item 16, above), makes clear in Art. 3 that, for the purpose of the Protocol:

(a) "Trafficking in persons" shall mean the recruitment, transportation, transfer, harboring or receipt of persons, by means of the threat or use of force or other forms of coercion, of abduction, of fraud, of deception, of the abuse of power or of a position of vulnerability or of the giving or receiving of payments or benefits to achieve the consent of a person having control over another person, for the purpose of exploitation. Exploitation shall include, at a minimum, the exploitation of the prostitution of others or other forms of sexual exploitation, forced labor or services, slavery or practices similar to slavery, servitude, or the removal of organs.
(b) The consent of a victim of trafficking in persons to the intended exploitation set forth in subparagraph (a) of this article shall be irrelevant where any of the means set forth in subparagraph (a) have been used.
(c) The recruitment, transportation, transfer, harboring, or receipt of a child for the purpose of exploitation shall be considered "trafficking in persons" even if this does not involve any of the means set forth in subparagraph (a) of this article.
(d) "Child" shall mean any person under 18 years of age.

It is important to note that the victim's consent, or the consent from his or her legal representative, is irrelevant and does not exculpate the defendant from human trafficking charges under the Protocol. For a long time prior to the adoption of this Protocol, and especially in cases of crimes of a private nature, consent operated as an affirmative defense that led to acquittal from such crimes. According to Feuerbach, "while a person may waive his rights through a declaratory act of his will, consent of the fact from the injured party eliminates the concept of offense" ("volenti non fit injuria").[18] Therefore, a valid consent depends on the value the law seeks to protect. If the solution is restricted to principles of civil law, then the autonomy of criminal law is relinquished.

Zitelmann recognizes that the legal nature of the consent may explain its effectiveness as a defense to unlawful acts, despite the silence of the law on this issue. The principles governing legal businesses exist in private law and can justify the mentioned exclusion. Bierling and Mezger have adopted the "theory of legal action," which subscribes to the belief that these legal provisions are safeguarded so long as the holder considers them valuable. Welzel, who has also adopted the theory of legal action, reveals that consent must be serious and match the true will of the victim. He does not take into account consent that is given while incapacitated or obtained by coercion.[19]

Criminal codes in the past contained little about the validity of consent. For example, in Brazil, the "Criminal Code of the Empire" (1830) contained no provision regarding consent, while Art. 26 in the Criminal Code of 1890 stated that "Criminal intent is not excused by… c) The consent of the victim, except in cases where the law permits it." Although Art. 14 in the Alcântara Machado Bill mentioned the consent of the victim,[20] the Revision Commission of the 1940 Code, led by Nelson Hungria, decided to delete the provision on the grounds that it would be superfluous.

Examining different criminal laws today, numerous laws exist where consent is a part of the typical structure of the crime or is presented as an expressed element (e.g., Art. 150 of the Brazilian Penal Code, trespass of one's home) or as a tacit element (e.g., Art. 151 of the Brazilian Penal Code, opening mail addressed to another, and Art. 153, disclosing contents of private documents). Under these laws, consent works as a way to negate the offense itself. José Henrique Pierangeli divides such crimes into four groups: (1) crimes against property or patrimony, (2) offenses against the physical integrity, (3) offenses against honor, and (4) crimes against individual freedom.[21]

---

[18] PIERANGELLI, José Enrique; in: *O Consentimento do Ofendido na Teoria do Delito*. São Paulo: Revista dos Tribunais, 1989, pp. 68–75.

[19] Cf. *id.*

[20] Art. 14: "There is no punishment for those who perform the crime: I—with the consent from whom is able to waive the violated or threatened right." The Alcântara Machado Bill relates to the Brazilian Penal Code.

[21] *In*: PIERANGELLI, José Henrique. *O consentimento do Ofendido na Teoria do Delito*. São Paulo: Revista dos Tribunais, 1989, p. 93.

The implications of consent become problematic in the context of financial crimes, where the sole holder of the protected right who can prevent the crime is the same person who acquiesces and thus provides consent, thereby preventing prosecution of the crime. However, consent carries no legal validity with respect to offenses that necessarily cannot include consent, such as theft or robbery. For instance, for crimes against physical integrity, the Brazilian legislation does not contain any provision regarding consent. While suicide and self-inflicted harm are not punished in and of themselves, self-inflicted harm can be punished if it is used as a means to achieve a criminal end. However, for minor assaults, the doctrine diverges and at times considers the victims' consent to the assault.[22] In the case of offenses against honor, consent may be used as a justification for the offense; the same can be said with respect to crimes against individual freedom that specifically involve sexual freedom.

It is also important to outline an analysis of consent from the viewpoint of the "objective attribution theory." With respect to material offenses, it is commonly understood that the person who caused the offense to occur should be the one prosecuted. This is why criminal law in particular focuses on the relationship between the act committed by the suspect or defendant and the produced result of that act. This focus on causation has existed for some time, whereby the attribution of the resulting crime to its rightful instigator and attaching criminal liability to this person are considered to be of utmost importance. Thus, it is important to verify whether the causal relationship between the act and the resulting crime was sufficient to assign responsibility for the crime to the perpetrator of the act.

Juan Bustos Ramirez makes it clear that causation should be the exclusive criterion considered with respect to a crime, relegating any discussion regarding imputation to secondary status.[23] Nelson Hungria states that analyzing the causal relationship is indispensable when charging someone with a crime.[24] Based on this understanding, many penal codes have adopted the "but for causation theory," or *conditio sine qua non,* for determining causation. For example, the Brazilian Penal Code says in Art. 13: "the existence of crime is attributable only to someone who caused it." In order to determine whether an act is a cause of the resulting crime, one option is to use the method proposed by Thyrén, where one hypothetically eliminates the act and determines whether the result would have occurred anyway. If so, the causal link does not exist and the act should not be prosecuted.

Extensive and slightly different discussions regarding causation have also begun in Germany, where scholars reflect on whether the causal theory is sufficient to attribute a crime to its instigator or if it is possible to attribute responsibility to an actor who may not have directly caused the crime, but could foresee its occurrence. Claus Roxin resizes the issue, perfecting what Richard Honig (*Causality and Objective Imputation*) and Karl Larenz (*The Attribution Theory in Hegel and the*

---

[22] *In*: *Lições de Direito Penal*—parte especial, p. 92.
[23] Cf. LARRAURI, Elena. *La imputación objetiva*. Santa Fé de Bogotá/Colômbia: Temis, 1998., pp. 7–8.
[24] *In*: *Comentários ao Código Penal*, arts. 1 a 27, p. 197.

Concept of Objective Imputation: Methodology of Science of Law) began to explain last century, by stating that what is relevant in criminal law is not whether a certain event occurs, but rather whether it is possible to examine the criteria and attach culpability for the resulting crime to the suspect who committed a certain act.

This shift in perspective from but-for causation to foreseeability occurs while analyzing the conduct itself, before considering the subjective intent or fault behind the conduct. This shift does not seem to occur with respect to other aspects of the illicit activity, such as defenses that negate liability. Hence, the issue of foreseeability requires an objective analysis that ultimately determines whether the suspect should be prosecuted for his or her acts. This explains why this mode of analysis is named the "objective attribution theory."

Roxin originally conceived the idea of analyzing culpability solely within the context of the legality of one's actions. Based on this context, it was very important to determine whether the conduct and its consequences fell within the acceptable boundaries of the law, rather than analyzing the relationship between the conduct and the consequence. The conduct and its outcome must be driven by an unlawful or unfair intent. This view is echoed by José Carlos Gobbis Pagliuca, who states that the objective attribution theory "selects conducts to be the focus of Criminal Law only those that are considered reprehensible due to the behavior of the perpetrator and the risk to the legal interest."[25]

As an example that illustrates this point of view, Roxin discusses the death of a victim during a lightning storm. Could such a result be attributed to the person who sent the victim into a hill with tall trees, hoping that the victim would get struck by lightning and that the suspect would therefore receive an inheritance? For the renowned German professor, this would not occur because the conduct alone is not unlawful, and as such, is not a candidate for culpability. The same result is obtained when a person dies due to a traffic accident that occurs while being taken to the hospital: The person who caused the initial accident cannot be held criminally liable for the ultimate death in the ambulance, even under manslaughter charges.

However, there may not be prohibited conduct if a decrease or an alleviation of a relevant legal risk occurs during the perpetration of hazardous conduct. A decrease of risk occurs if, for example, someone throws a stone believing it will fall on the head of a victim, but it hits a less vulnerable part of the body instead. In this case, Roxin pontificates that there is a "reduction of risk in relation to the protected legal interest and, therefore, it is not to characterize the fact as a criminal offense. The conduct which reduces the likelihood of injury cannot be conceived as oriented in accordance with the purpose of damaging the physical integrity." This thinking is considered valid for all cases that involve mitigating events which decrease the harm to the victim.[26]

---

[25] Cf. PAGLIUCA, José Carlos Gobbis. A imputação objetiva (quase) sem seus mistérios. *Revista da Associação Paulista do Ministério Público*. São Paulo: ano IV, n. 35, out./nov./00. p. 35.

[26] *In*: ROXIN, Claus. *Problemas fundamentais de direito penal*. Trad. Ana Paula dos Santos Luís Natscheradetz (Textos I, II, III, IV, V, VI, VII e VIII), Maria Fernanda Palma (Texto IX) e Ana

On the other hand, certain activities, due to their nature and social utility, are allowed to maintain their hazardous elements, thus allowing the increased risk of harm to others. For such activities, it is not always possible to place culpability on the direct cause of the harm. As an example, Roxin cites the case of a manufacturer of brushes made from goat hair who violates regulations by improperly using disinfecting material to manufacture the brushes. As a result, four workers die after contracting an infection due to exposure to anthrax bacilli. In such a case, those deaths are attributed to the employer, even though proper disinfection cannot guarantee absolute destruction of the bacilli. The regulations are written such that those employers who do not comply with the regulations assume all the liability for such a risk. If the risk was considered to be very high, the law would completely ban the use of goat hair entirely.[27] In this case, there was evidence of increased risk caused by failure to comply with regulations, thus making the employer criminally liable for the resulting harm.

Finally, the commission of hazardous conduct may still be subjected to limited liability when the harm that occurs is not within the scope of the harm contemplated by those who prohibited this particular conduct. According to Hans-Heinrich Jescheck, when "the result stands outside the scope of the law that the perpetrator has offended with his actions, the legally disapproved risk that the perpetrator created does not materialize in the outcome, but a different risk does."[28] For example, when a driver who hits a pedestrian is driving with an expired license or is carrying alcohol in a prohibited manner in his car, there is a clear lack of connection between the scope of the law that was violated and the subsequent accident.[29]

The merit of the objective attribution theory is that it raises new considerations with respect to causation, although these considerations may not be as relevant within the context of material crimes. Significantly, this theory does not construe conduct using a simple cause-and-effect analysis, but rather focuses on whether the conduct violates criminal laws and is aimed at achieving illicit purposes. This theory justifies the prosecution of criminal actions or omissions that do not depend on reaching a particular result. The punishment of such conduct should not be based on whether it caused a certain outcome, but rather, as Welzel properly explains, it should be based on the means that were used and their relationship to the law. He believes that "the injustice of crimes without intention is based on the discrepancy between the action actually undertaken and the one that should have been made with care."[30]

Isabel de Figueiredo (Texto X). 3. ed. Lisboa: Vega Universidade; Direito e Ciência Jurídica, 1998, p. 149.

[27] Cf. *id.*, p. 153.

[28] Cf. JESCHECK, Hans-Heinrich. *Tratado de derecho penal: parte general.* Trad. José Luis Manzanares Samaniego. 4. ed. Granada: Comares, 1993, p. 259.

[29] Cf. ROXIN, Claus. *Op. cit,* p 155.

[30] Citado por BITENCOURT, Cezar Roberto. *Manual de direito penal: parte geral.* 6. ed. São Paulo: Saraiva, 2000. v. 1. p. 156).

For crimes that are often perpetrated within the sports sector, whether the victim consented does not seem relevant enough to minimize the resulting culpability, even under the objective attribution theory. Indeed, Brazilian law includes stringent punishments for crimes that other jurisdictions have not prosecuted as forcefully, especially with respect to crimes related to forced labor. It is clear why mere consent cannot negate culpability for crimes that involve the exploitation of human beings, since such crimes are predicated on obtaining the victim's consent through various means, including promises of economic well-being. For example, Art. 3 of the Protocol to Prevent, Suppress and Punish Human Trafficking, Especially Women and Children states that, for purposes of the Protocol, "human trafficking" requires the use of deception or coercion to take advantage of the vulnerability of victims, including using the promise of payment or benefits to obtain the victims' consent.

Given the existence of these protections, it is egregious that football agents are still able to take advantage of the financial and moral frailty of young players in an effort to exploit them. In order to recognize suspicious circumstances involving young people within the football sector, it is important to understand that even in case of a "strong" victim—i.e., one who has freely acquiesced to his situation without being deceived or coerced—it is impossible to eliminate the criminal liability that accompanies such acts.

Despite the assertions supported by the objective attribution theory, what is at stake when we allow limited culpability for such crimes is the effective protection of human dignity. Human dignity is a universal principle, and an offense against an individual's dignity that goes unpunished weakens the preservation of human dignity as a whole. It is not possible to speak about criminal practices within the world of football without acknowledging the vulnerability of the victims, even if this vulnerability is not related to the players' physical or emotional strength. This vulnerability should presumptively exist under the law. As discussed above, the victim can easily be deceived or coerced by the offender based on the promise of obtaining a better life. Based on the obvious emotional and economic exploitation that often occurs within this sector, the presumption of vulnerability should not be easily overcome, thus allowing increased legal protection against such acts performed by football agents.

This need to consider the vulnerability of the victim on a case-by-case basis seems to be the focal point of recently passed legislation. For instance, the Working Committees for the study of the Brazilian Civil Code, in a conference held in Brasilia and organized by the Council of the Federal Court in September 2002, culminated in revisions to Statement 23, which considered the content of Art. 421. The revisions emphasized the prevalence of human dignity within contractual autonomy and demonstrated concern, even in the field of private law, about the inherent value of all individuals.

José de Faria Costa and Manuel da Costa Andrade emphasize that "economic crime is refractory to an objective valuation of the worthlessness of the result—this often translates into only a breach of trust in the economic system. Delmas-Marty emphasizes this as well—congressmen feel the need to consider that the mere endangering certain legal rights is an element of crimes (concrete danger) or even the risk is not a crime element (abstract danger). It is clear that abstract danger crimes greatly facilitate the overcoming of difficulties of evidential proof in economic crimes, which led many countries (Germany, Austria, Belgium, Spain, France, Greece, Japan, Poland, and Switzerland) to use this form of crime."[31]

Klaus Tiedemann recognizes that abstract danger crimes constitute a considerable restriction of the freedom of entrepreneurial action, but reasons that the prohibition to perform certain actions corresponds to the idea that the law cannot prevent the production of harmful results, but only the carrying out of dangerous practices. Tiedemann also points out that criminal prosecutions involve less severe interference than administrative law investigations, where an entrepreneur's overall activity can be submitted to surveillance by the state. Criminal prosecutions only focus on the activity related to the specific offense.[32]

The rise of transnational crime has triggered a reflection on the effectiveness of traditional criminal law to suppress it. For example, Mireille Delmas-Marty and Geneviève Giudicelli-Delage assert that the UN Convention held on December 20, 1988, which focused on drug trafficking, represented nothing more than the failure of national systems to cope with crime, and that "beginning in the late 1980s, the international community became aware of the shortcomings—if not futility—of national rights when faced with increasingly effective international crime prospering precisely because of the disparities between, and lack of harmony among, national legislative bodies.... The UN Convention signed at Vienna on December 20, 1988, was the first response to bring harmony to enforcement."[33]

Parallel to such discussions, the national governments were obliged to enact increasingly severe laws, trying to embrace this new form of criminal activity without borders. However, in response, more sophisticated methods and techniques are used to wipe out all vestiges of crime—methods that mutate in response to changes in crime-fighting techniques. This calls for nimble, flexible legislation that is able to combat the realities of economic crime effectively. Therefore, it is important to create effective tools to combat crimes within the sports sector, breaking the cycle of ill-gotten gains by not restricting such crimes to an exhaustive list of offenses. Moreover, the analysis must consider individual rights and guarantees, but it should also consider the vulnerability and human dignity of these victims as a whole.

---

[31] Sobre a concepção e os princípios do Direito Penal Económico. *Direito Penal Económico e Europeu: Textos Doutrinários*, vol. I., p. 356.

[32] Cf. TIEDEMANN, Klaus. *Poder económico y delito* (Introducción al derecho penal económico y de la empresa). Barcelona: Ariel, pp. 33–34. 1985.

[33] "C'est à la fin des années 1980 que la communauté internationale prend conscience de l'inadéquation—et même de l'inanité—des droits nationaux em face d'une criminalité d'autant plus efficace qu'elle se déploie à l'échelle internationale et prospère précisément sur le terreau de la disparité et de l'absence d'harmonisation des législations nationales (...) La Convention ONU, signée à Vienne le 20 décembre 1988, traduit la première ce souci d'harmonisation." *In: Droit pénal des affaires*, pp. 309–310.

Bibliography

ARRUDA, Eduardo. A ameaça. FOLHA DE SÃO PAULO, ESPORTE, D2, PAINEL FC, March 31, 2010.
BITENCOURT, Cezar Roberto. Manual de direito penal: parte geral. 6. ed. São Paulo: Saraiva, 2000. v. 1.
CAPEZ, Fernando. Curso de Direito Penal—parte especial. São Paulo: Saraiva, 2008.
COLON, Leandro; RIZZO, Marcel. Havelange, 96, renuncia a cargo na Fifa para não sofrer punição. Folha de São Paulo, Esporte, D4, May 1, 2013.
CORREA, Luís Fernando. A história do doping nos esportes. In: http://g1.globo.com/Noticias/Ciencia/0,,MUL1267929-5603,00-A+HISTORIA+DO+DOPING+NOS+ESPORTES.html, August 16, 2009, accessed June 9, 2013.
CORREIA, Eduardo. Introdução ao direito penal econômico. In: CORREIA, Eduardo et al. Direito penal econômico e europeu: textos doutrinários. Vol. 1. Coimbra: Coimbra Ed., 1998. pp. 293–318.
CORREIA, Eduardo. Introdução ao direito penal econômico. Revista de Direito e Economia, no. 3, 1977.
CORREIA, Eduardo. Novas críticas à penalização de atividades econômicas. In: CORREIA, Eduardo et al. Direito penal econômico e europeu: textos doutrinários. Vol. 1. Coimbra: Coimbra Ed., 1998. pp. 365–373.
DELMAS-MARTY, Mireille, Sobre a concepção e os princípios do Direito Penal Económico. Direito Penal Económico e Europeu: Textos Doutrinários, vol. I.
DELMAS-MARTY, Mireille. Droit penal des affaires. 3. ed. Paris: Presses Universitaire de France, 1990, t. 1.
DELMAS-MARTY, Mireille; GIUDICELLI-DELAGE, Geneviève. Droit penal dês affaires. 4 ed. Paris: Presses Universitaire de France, 2000.
FIFA Code of Ethics, http://www.fifa.com/aboutfifa/organisation/footballgovernance/codeethics.html, accessed June 1, 2013.
FIFA Regulations on the Status and Transfer of Players, http://www.fifa.com/mm/document/affederation/administration/01/95/83/85//regulationsstatusandtransfer_e.pdf, accessed June 21, 2013.
FRAGOSO, Heleno Cláudio. Lições de direito penal: a nova parte geral, 8ª ed. Rio de Janeiro: Forense, 1985.
GREENFIELD, Steve; OSBORN, Guy. The Football (Offences and Disorder) Act 1999: Amending s3 of the Football Offences Act 1991. J. C. L. 55 2000, p. 62, accessed April 1, 2013.
HUNGRIA, Nelson. Comentários ao Código Penal: arts. 1 a 10, 11 a 27, 75 a 101. 4. ed. Rio de Janeiro: Forense, 1958.
IRVING, James G. Red Card: The Battle Over European Football's Transfer System. 56 U. Miami L. Ver. 667, 2001.
JESCHECK, Hans-Heinrich. Tratado de derecho penal—Parte general. 4th ed. Transl. José Luis Manzanares Samaniego. Granada: Comares, 1993.
JOHNSON, Graham. Football and Gangsters. How organized crime controls the beautiful game. Great Britain: Cox and Wyman Ltd., 2007.
LARRAURI, Elena. La imputación objetiva. Santa Fé de Bogotá/Colômbia: Temis, 1998.
NELSON, Timothy G. Flag on the Play: The Ineffectiveness of Athlete-Agent Laws and Regulations—and How North Carolina Can Take Advantage of a Scandal To Be A Model For Reform. 90 N.C. L. Rev. 800, 2011–2012, p. 803, April 1, 2013.
PAGLIUCA, José Carlos Gobbis. A imputação objetiva (quase) sem seus mistérios. Revista da Associação Paulista do Ministério Público. São Paulo: ano IV, n. 35, out./nov./00. p. 35.
PIERANGELLI, José Enrique; in: O Consentimento do Ofendido na Teoria do Delito. São Paulo: Revista dos Tribunais, 1989.
RACISMO poderá gerar até rebaixamento. FIFA. Federação cria pacote de propostas para combater discriminação, com punições financeiras e esportivas. Folha de São Paulo, esporte, D3, May 7, 2013.

ROXIN, Claus. *Problemas fundamentais de direito penal*. 3rd ed. Transl. Ana Paula dos Santos Luís Natscheradetz (Textos I, II, III, IV, V, VI, VII & VIII), Maria Fernanda Palma (Texto IX) & Ana Isabel de Figueiredo (Texto X). Lisbon: Vega Universidade/Direito e Ciência Jurídica, 1998.

ROXIN, Claus. *Funcionalismo e imputação objetiva no direito penal*. Translation and Introduction by Luís Greco. Rio de Janeiro: Renovar, 2002.

ROXIN, Claus. *Derecho penal—Parte general—Fundamentos. La estructura de la teoría del delito*. Vol. I. Madrid: Civitas, 2006.

ROXIN, Claus. *La teoria del delito en la discusión actual*. Transl. Manuel Abanto Vásquez. Lima: Grijley, 2007.

ROXIN, Claus. Reflexões sobre a construção sistemática do direito penal. *Revista Brasileira de Ciências Criminais*. Vol. 82. pp. 24–47. São Paulo: Ed. RT, 2010.

SIEKMANN, Robert. Labour Law, the Provision of Services, Transfer Rights and Social Dialogue in Professional Football in Europe. 4 ESLJ 1 2006–2007.

TIEDEMANN, Klaus. Poder económico y delito (Introducción al derecho penal económico y de la empresa). Barcelona: Ariel, pp. 33–34. 1985.

Chapter 3
Football

3.1   Football: A Dream or a Nightmare?

Football, one of the world's most well-liked and celebrated sports, is the subject of much attention and involves many foreign authorities. Football is practiced by more than 265 million people in the world. According to FIFA, there are 38 million registered professional players and approximately 301,000 clubs.

The extraordinary growth of football began in the early 1990s as a result of increased television rights and sponsorship. The market for professional players has experienced unprecedented internationalization, which has led to more transfers of resources by transcontinental dimensions. This sharp increase in economic and international development makes football more susceptible to organized crime. This danger cannot be ignored by state controls based on social, educational, and cultural good practices.

The distribution of football has led to a hugely disproportionate drain on resources. According to Natalie L. Clarke St. Cyr, "the top 20 clubs alone generated around $ 6 billion in the 2009–2010 season. Despite increasingly high volumes of revenue, European football clubs are finding themselves saddled with large amounts of debt."[1] Football revenues come from ticket sales, television rights, sponsorships, and other commercial activities, such as selling merchandise. The majority of this revenue is used to pay the players' salaries and cover the costs of their transfers.

Football is an attractive area to conduct financial crimes because it is an easy market to penetrate. The football industry consists of a network of mediators, managers, and players. Several actors in the football industry affect the sector financially: clubs, players (the most valuable asset to the clubs), sponsors, individual investors (patrons and club owners), football agents (acting in the interest of the players or as the intermediary in charge of transfers of athletes), and stadium owners. This structural diversity creates a lot of moving parts that can easily hide illicit activity, especially because this structure incorporates the international market. Moreover, the movement of large amounts of money, the difficulty in accounting for all these

---

[1] In The Beauty and The Beast: Taming the Ugly Side of The People's Game. 17 Colum. J. Eur. L 601, 2010–2011, p. 602.

F. M. De Sanctis, *Football, Gambling, and Money Laundering*,
DOI 10.1007/978-3-319-05609-8_3, © Springer International Publishing Switzerland 2014

transactions, and, ironically, the clubs' own financial needs increase this sector's vulnerability to crime.

It is also important to note the vulnerability of the athletes involved. These players and their families, many of who come from circumstances of poverty, think of football as a solution to their problems. This perception of the sport makes them reluctant to acknowledge the existence of corruption within the sector or to respond to criminal acts that occur. Many who call themselves "entrepreneurs" take advantage of this perception by defrauding poor families. There are reports of people pretending to be experienced European club entrepreneurs who promise wages for children whom they have never even seen playing football. In exchange, they charge the families € 10,000 per child. The families, duped by false documents that seem to authenticate the legitimacy of these entrepreneurs, sell their homes to pay for the required fees of these phony businesspersons. Thus, these con artists cause the families to lose everything they own in the hopes of seeing their child become an international player, a hope that will never be realized.

It is estimated that approximately 20,000 boys live on the streets in different parts of Europe as a result of these scams.[2] The vast majority of these young players have been deceived by these so-called entrepreneurs who brag about their personal influence in European football and promise a brilliant future for these boys. Often, these boys do not even reach Europe and are rescued from small boats, where they are exposed to a high risk of harm or even death. Even those who manage to reach the European continent usually end up discovering that there is no way to contact the clubs because these clubs, in fact, do not have any interest in them. This leads to an increase in the number of illiterate, abandoned children, left with no shelter, no visa, no identification card, and no money. In addition, these children typically cannot even speak the language of the country where they are stranded.

Moreover, for those players who are legitimately called to participate in these clubs, the irony of the situation is how often the "dream" of football turns into a nightmare for these players. Many will work for 3 or 4 months and receive insufficient wages. Uncertain about their future, players will travel thousands of miles using precarious transportation and sleep in cramped quarters with dozens of men sharing a few rooms. Intending to provide for their families, many are often forced to leave their families, including their spouses and children, in order to follow this "dream."

In one particular club, for example, nearly 30 athletes would sleep in the same 70-square-meter shed in 15 bunks, with no cooling system in a climate that often reaches 35 °C (100 °F). After training, the players themselves had to heat their food and clean the bathrooms, and the kitchen in these accommodations were located approximately 15 km away in a mosquito-infested area. It was not unusual for players to contract diseases from the mosquitos, forcing these players to return to their hometowns due to illness. Moreover, these players had to drink water drawn from

---

[2] Cf. JOHANSSON, Jens; MADSEN, Lars Backe. On the "muscle drain" and (child) trafficking and football. *Den Forsvunne Diamanten,* Tiden Norsk Forlag, Noruega, October 2008.

an artesian well, and the lack of familiarity with local, unpurified water caused many players to contract diseases from various bacteria and viruses.

There are several other cases involving boys between the ages of 13 and 15 in different regions of Brazil. For example, players lived in abandoned buildings in Bangu City, west of Rio de Janeiro. The buildings often had an insufficient number of beds, the food was spoiled, and the young players were exposed to mold, leaks, and fly larvae. Moreover, the alleged transfer agent was not licensed by FIFA.

These players are often forced to remain in these horrible circumstances because they unknowingly signed 6-month contracts. These short-term contracts, combined with the lack of proper wages, prevent players from visiting their relatives, spouses, or children, many of whom are several miles away. The difficulty in accessing Internet creates another barrier between the players and their loved ones, accentuating the distance between them.[3] Football agents and clubs use these adhesion contracts as a cruel marketing tactic, aimed at recruiting athletes from several places, housing them in substandard conditions, and then abruptly dismissing them when their tournaments end. Within the first week of March 2010, approximately 200 contracts that were signed earlier that year had already been brought to an abrupt end.

Despite the meager pay and even dangerous conditions that players must often withstand, many athletes and would-be players still support the football industry. The desire to fulfill their dreams of winning games and making it big overcomes any fear they may feel as they wait for a position within a club. Ironically, in Brazil, 90 % of jobs in football do not even recruit players based on their performance in the clubs. Instead, they recruit based on whether the player has a special connection to a particular coach.

There are instances when certain players are fortunate and able to get out of improper or unsafe conditions. For instance, a 21-year-old player from a famous club in Rio de Janeiro received a proposal to play football in Russia by three businessmen—a Brazilian, a German, and an Italian—during an excellent time in his career. Upon his arrival, the businessmen wanted to submit the athlete for training, which he promptly refused. These businessmen subsequently broke their promise to the player by attempting to pay him only half the amount he had negotiated, without the right to a subscription premium. The player's mother notified the Brazilian Consulate, and with its help, the player was able to return to his country.

There are other cases in which Brazilians have given up football after experiencing even more intense harms. For example, a player from a major club in Minas Gerais went to train with a large club in Spain, where he shared a room with many other boys. After 28 days in Spain, the player still had not eaten. He eventually returned to Brazil with the help of his parents.

Outside Brazil, the reports are not very different.[4] In one case, an entrepreneur in charge of a first-division Norwegian club imported athletes from Nigeria to Oslo

---

[3] These cases were mentioned in a newspaper article: Em Rondônia, concentração dura até a bola parar. Folha de São Paulo, Esporte, D11, 14 Mar. 2010.

[4] See Money laundering through the football sector, available at http://www.fatf-gafi.org/media/fatf/documents/reports/ML%20through%20the%20Football%20Sector.pdf, accessed Feb 4, 2014.

without paying for them. The Norwegian entrepreneur connected the Nigerian players to an important and well-known English club via a top-secret scheme, where the Norwegian club would be paid to hide the minors until they reached the age of 18 and could be brought to England legally. However, the Norwegian entrepreneur also decided to sell the most talented Nigerian players to another English club upon reaching the legal age to play, hoping that the first English club would not sue based on this breach of contract due to the illegal nature of the business with the Norwegian club.

Contrary to what the Norwegian entrepreneur predicted, agents of the first English club arrived in Norway to reclaim the players sold to the second English club. This caused the second English club to write to FIFA, asking the organization to ban the Norwegian entrepreneur and exclude the first English club from the Champions League and from participation in the player transfer market. The issue was ultimately resolved through a confidential agreement that required a payment of 20 million pounds to the first English club—12 million from the second English club and 8 million from the Norwegian entrepreneur. The first English club was allowed to keep its Nigerian players. Due to the confidential nature of the negotiations, FIFA never became aware of the agreement, and no punishment was ever enforced against any of these three entities.

Cases like the ones discussed here reveal the existence of fraudulent activity and dangerous conditions that subject players to a kind of "consensual" slavery in football. This is a highly urgent problem, not only because it negatively impacts several young players and their families but also because con artist "entrepreneurs" compromise the work of legitimate football agents, thus tarnishing the credibility of the sector as a whole.

3.2 Typologies (Money Laundering Methods)

The various methods used to launder money in football can be summarized using the following categories, which are elaborated upon in the Financial Action Task Force (FATF) discussions: (1) acquisitions and investments in football clubs, (2) the international market for the transfer of players, (3) the "acquisition" of players, (4) selling tickets to games for unlawful purposes, (5) betting, (6) misuse of image rights, and (7) sponsorship and advertising.

With respect to investments in clubs, there are several cases reported by the FATF where illicit money was laundered, while maintaining the semblance of a legitimate investment going through the regular economic channels. An interesting example of this phenomenon occurred in France. After receiving a suspicious transaction report from an accountant, the financial intelligence unit (FIU) investigated a certain club's accounts and found that these accounts had always been in deficit, but gave the impression of being in balance through payments from one entrepreneur and his companies. The FIU discovered that the entrepreneur was also the owner of the club, and the investments were made without any real compensation. The investments in the club led to his companies incurring losses, since the amount invested in the club did not yield adequate returns. The FIU also found improprieties in the accounting records for these investments. For all these reasons, the case was submitted to the authorities to investigate whether there was improper use of corporate assets to launder money.

Investments in football can lead to great profit and a more favorable status within the sport sector. However, for the most part, it is rarely easy to determine the source of these football investments. The murkiness of the transactions makes it extremely difficult to verify the origins of the funds, thus making it easy to hide any evidence of dirty money within the transaction. Such investments are made in both amateur clubs and more established ones. It is interesting that, despite the number of large commercial contracts that produce a great amount of revenue for the established clubs, these clubs still find themselves in precarious financial situations that require an increasing number of resources, i.e., more investments.

Reports have also surfaced regarding investments in football made by drug traffickers. One case reported in Mexico involved a person from the countryside who moved to the coast and subsequently obtained a vast amount of money, which he used to set up several companies and projects. One of his projects included a third-division football team that had no reason to attract such investments. The investments made in this football team paid for the high salaries and infrastructure costs, which ultimately allowed the team to upgrade to second-division status. While this higher status gave some legitimacy to the investments, it was later discovered that the origin of the investments was drug trafficking money, specifically from the drug trafficking network headed by the person who started the football team.

Known as "emotional laundering," an increasing number of cases have emerged where investments are made in football clubs without a screening taking place to determine the origin of the funds. Investments that are made to acquire clubs can be reversed through fictitious loans that enable illicit monetary gains. The investor, through an agent located in an offshore tax haven, buys players, while the seller company pays fees through this agent. The value of the fees is similar to the value of the investments. Such fees are passed on to other offshore accounts controlled by the original investor. To avoid such methods of money laundering, it is important to obtain appropriate documentation and further information in order to identify the people, the transactions involved, and the source of funds for each club or player. It is also important to recognize the real beneficiary and the real controllers of the club.

With respect to player transfers, their vulnerability to money laundering has increased as the market for transfers has broadened. The internationalization of the transfer market began in the 1970s, when Italy, England, and Spain increased the possibility of hiring foreign players in 1974, 1978, and 1980, respectively.[5] Over the

---

[5] Furthermore, according to statistical data compiled by the Brazilian Financial Intelligence Unit (COAF), the number of Suspicious Activity Reports has been very low—in 2012, there were 23, in 2013, 86 (including expositions, artists, and athletes), which shows that the law is not very effective (in COAF, https://www.coaf.fazenda.gov.br/conteudo/estatisticas/comunicacoes-recebidas-por-segmento, accessed Jan. 22, 2014).

years, the best players from all over the world have been hired by various European countries. Brazil has been a major exporter of athletes, with some of their athletes even transferring to Asia. Meanwhile, Africa is still considered a cheap source of new talent.

In France, a suspicious transaction report involving player transfers led to the discovery of illicit activity conducted by an individual whose company specialized in consulting and advising people connected to the sport sector. This person later admitted to being a clandestine agent for players. Large amounts of money were sent through wire transfers to the suspect's company, and it was later found that most of the credits of his company originated from another company belonging to an Eastern European known for his ties to organized crime in his country. Information collected by the FIU revealed that these two entrepreneurs operated their businesses with the goal of facilitating money laundering.

FIFA has tried to restrict the activities of clandestine agents to prevent the commission of crimes within the football sector. These restrictions should also be applied within the territories of the football associations or confederations. André Dinis Carvalho states that "When the sport entrepreneur emerged, clubs often denied negotiating with them, who were considered persona non grata that seek to maintain the superiority in the process of negotiation."[6]

It is imperative to regulate and supervise the performance of these agents, lest they storm the market and support interests that are rooted in the economic exploitation of the sport sector and parallel businesses that revolve around the sport industry. However, this oversight is currently not occurring. Many agents still operate without a regular license, and athletes purposely choose such agents to get around official regulations and controls. The existence of these agents makes the sports sector less transparent and more difficult to monitor. In addition, agents (licensed or not) have formed a closed community, which makes it difficult to verify transactions conducted by them.

FIFA adopted a Regulation on the Players' Agents on May 20, 1994, which was then modified in December 1995, revised again on October 29, 2007, and has been made effective since January 1, 2008. This document clarifies that the FIFA agent would be empowered, only through obtaining a license, to care and to act on behalf of players and/or clubs for player transfers anywhere in the world. On October 12, 2000, the FIFA Executive Committee approved new Players' Agents Regulations,[7] which became effective on January 3, 2001, to fit the requirements of the European Parliament. Under these regulations, any resident of the European Union or the European Economic Area may request an agent license from a national federation in any country where he or she was born and has legal domicile and may also get liability insurance from a company in any of these countries. With the new edition

---

[6] In A profissão de empresário desportivo—uma lei simplista para uma actividade complexa? Desporto e Direito, Revista Jurídica do Desporto, Ano I, n. 2. Coimbra: Coimbra Editora, Jan./Apr. 2004, p. 253.

[7] FIFA Regulations Players' Agents. In http://www.fifa.com/mm/document/affederation/administration/51/55/18/players_agents_regulations_2008.pdf, accessed July 1, 2013.

of the FIFA Players' Agents rules mentioned above, each national association can draw up its own rules, as long as they are based on FIFA rules and are approved by the FIFA Players' Committee. Moreover, the Committee set a time limit of 5 years on the duration of the FIFA agent license, requiring further assessment of the agent for renewal.

However, these FIFA regulations alone are not enough to deal with this issue. These regulations do not inhibit the conduct of opportunists who do not care about athletes and focus exclusively on speculative investments that often violate legal and ethical principles. The FATF makes clear that the physical movement of players is important to monitor, particularly because the circumstances surrounding the recruitment and transfer of these players are not always clear. Therefore, it is necessary to impose legal limits that avoid deleterious effects in the trading of athletes. Those who broker contracts for football players should register with the Central Bank and/or the Internal Revenue Service (IRS) in the player's country of origin. Moreover, there should be a prohibition on the inclusion of irrevocability clauses in labor representation contracts for children under 18 involved in sport, limiting the percentage perceived as fees to a finite and low number (e.g., 10 or 15%).[8]

FIFA regulations regarding the transfers of players do not cover all the financial transactions involved with respect to the transfer of a particular athlete from one country to another. For example, the national associations of each country are each responsible for presenting the International Transfer Certificate to FIFA. However, this certificate does not allow verification of monetary details regarding the international transfer. Therefore, there is no way to obtain useful information about accurate financial flows among clubs involved in such contracts.[9] This is to the detriment of regulating fraudulent transactions—many complex transactions occur during such transfers, and this along with the lack of communication with FIFA makes these international transfers more vulnerable to money laundering.

People not linked to football players are still able to acquire rights over them by using offshore accounts with complex and impenetrable financial structuring. Thus, a significant sum of money is usually obtained as a result of conducting transactions involving players. Investors sometimes invest in young, talented athletes, and later return a profit when these players are sold to another club. Money laundering can occur during the course of this sequence. For example, a club buys a player for ten and states that investors only invest half of the amount, since the difference is coming from dirty money. The player is later resold for 15 and everyone involved will get a profit based on this purchase and sale of the player: the club gets five, while the investors who laundered the remaining five get triple their investment, rather than simply doubling it.

---

[8] FILHO, Álvaro Melo mentioned by Felipe Legrazie Ezabella (In: *Agente FIFA e o Direito Civil Brasileiro*. São Paulo: Quartier Latin, 2010, p. 53–54).

[9] FIFA denied Spanish Courts to get a copy of the contract celebrated between the Brazilian football player Neymar and Barcelona club (in http://esporte.uol.com.br/futebol/ultimas-noticias/2014/02/05/fifa-se-nega-a-enviar-contrato-de-neymar-para-justica-da-espanha.htm, accessed Feb. 06, 2014).

With respect to the stadium tickets, money laundering occurs when there are no controls regarding the acquisition and sale of tickets, thus making their sale easily subject to manipulation. Clubs often declare, for instance, the sale of all or a large part of their tickets, but the associations or confederations do not properly monitor these sales to ensure that they are legitimate, especially when it comes to clubs in the lower divisions.

With respect to betting, placing bets on games can also be a form of money laundering. Many scandals in several countries revealed manipulations of game results for purposes of betting, seriously tarnishing the image of football. The International Police (INTERPOL) reported an operation in Asia that began in 2008 in which 1,300 people were arrested and US$ 16 million in cash was found. This money was suspected to stem from illegal betting operations on football games. This was not the first case of such organized crime—in 2007, another operation led to approximately 400 arrests, resulting in the seizure of US$ 680 million. This operation involved the use of gambling sites to launder money.

Meanwhile, in Belgium, it was found that many people went to casinos to purchase chips using large amounts of money. One person had no connection to the country where the gambling was happening, while others were directly related to football (i.e., players or coaches) or indirectly related (i.e., relatives of players or coaches). There was no valid economic justification for spending so much money at one time. It was later discovered that the payments were made to bribe players before certain games and to bet heavily on these games through an online gambling company. It was found that the "investments" were made in clubs experiencing financial difficulties that were seeking "rescuer" sponsors. The purpose was to make investments in these clubs as a way to control and bribe them into forgoing regulation against manipulated games. It was not until the resulting score of 8–0 in one of the games that authorities finally noticed the fraudulent conduct. Thus, the money used to buy these chips likely originated from criminal activity.

Unfortunately, there are no significant data about this type of money laundering, which apparently has the highest presence in Asia. However, this lack of data does not justify countries disregarding the existence of this form of criminal activity.

With respect to the possibility of committing crimes using image rights, sponsorships, and advertising, image contracts (i.e., contracts that allow for the use of a player's image or likeness for an extensive advertising campaign) are often used as a way to hide illicit money laundering. The money stipulated in such contracts is fraudulently transferred to accounts of companies in tax havens, even though there is no need for the use of third parties in these image contract transactions. Thus, the clubs experiencing financial need can accept cash payments under the guise of payments for image contracts. Players who are not as renowned can also agree to the use of their image.

The use of the image rights has also been a way to disguise laundered money as the salaries of players and their agents. For example, a player would have signed an agreement in England for the use of his image rights to a club and transfer money to a corporation located in a tax haven. In exchange, shares of the corporation would be given to the player. The club would have sought out an advisory professional who would have concluded that the player does not have a commercial interest in selling his image rights, thus preventing the club from exploiting the player's rights. However, through the corporation's involvement, renegotiation of such rights takes place, thereby increasing financial flows to the corporation. Indeed, the transaction would be part of the player's salary payments, which would allow the corporation to evade taxes while still benefiting from the use of the player's image.

Advertising and sponsorship contracts can also be used for money laundering purposes. Organized crime can sponsor clubs under the guise of conducting "legitimate" business. This sponsorship allows organized crime to acquire a dominant position within the sports sector by giving them access to the corporate network. Television broadcast rights are also a source of power within the sports sector, such that there are also confirmed cases of money laundering occurring via media contracts.

Besides the seven categories described above, it should be noted that rogue agents in various areas have established ways to use the sport sector to commit fraud and other financial crimes. For example, in Mexico, a businessman linked to the local government purchased a club, which would have been used as a way to attract politicians and public officials from various levels of government. Thanks to these contacts, it was possible for the businessman to reach out to people with the power to make decisions regarding public procurements, thus allowing the businessman to illegally obtain public contracts. In the UK, an international player's statement revealed that money arising from a contract he signed disguised payments to a foreign agent, allowing the club to evade social security payments.

3.3 Prevention and Control

Financial crime laws have become increasingly important in recent years. Many changes have taken place to keep up with the globalization of the economy. Until recently, there were great schisms between different parts of the world, such as the schism between capitalism and communist-driven socialism that sparked the Cold War. As the idea of a market economy advanced technological innovations, even in countries with no such tradition (such as China), there grew a need for new managerial practices specific to businesses.

Likewise, criminals evolved over time. Despite the positive hopes brought on by the advent of globalization, cutthroat competitiveness and destructive behaviors also developed. While we do not know where these behaviors will lead, the ominous trends that have already emerged have caused new and growing fears.

The inherent globalization in today's world, with all of its advantages and disadvantages, has fostered the technological and transnational criminal enterprise, often practiced by large businesses and conglomerates, that necessitates unprecedented cooperative exchanges among nations.

The special field of financial crime law is justified by the simple idea that marketplace rules alone cannot address all of the emerging aspirations within business practices that lead to the dangerous blurring of ethical lines. This is an area of criminal law designed to fill in loopholes in the definitions of crimes against property, owing, in large part, to increases in criminal infractions resulting from the exponential increase of economic activity within the state and of international financial relationships.[10]

The legal protection of property requires government intervention and social and economic regulation so that the rules of conduct with regard to business practices may be stabilized and preserved. Objects of legal protection enjoy this sort of global protection, not just by criminal law but also through the general expectation of stability fostered by proper norms and the honest functioning of markets consisting of corporations, private and public entities, and derivative securities.

Of course, general criminal law does apply, albeit in a fragmented, subsidiary way, as a last resort. In terms of financial crimes, the law will take on a substantial role rather than merely a symbolic role. Yet, when there are arguments based on the "needless" financial law often invoked indiscriminately, without the slightest basis in reason, the result is a systemic lack of protection of the economic order.

The takeaway here is that financial crime is a very current subject, "whether by the magnitude of the damage it causes, or by its capacity to adapt to, and survive, social and political changes, or even because of its readiness to come up with defenses and to defeat all efforts to combat it."[11] Conceptualizing financial crime is no simple task. It does not lend itself to simple measurement by the extent of resulting damages. The classification of financial crimes rests on the collectivized or supra-individual nature of the legal interests or assets that are to be protected.

Reducing intervention only to those alleged facts actually held meritorious is just as imperative as trimming away criminal liability hidebound by excessive formalism. Yet the fact remains that administrative sanctions alone have not sufficed to enforce the basic duties we, as citizens, are bound by as actors in the economic system.

Brazil's Federal Constitution, for example, went even further. Under the title *The Economic and Financial Order*, Chap. I, *The General Principles of Economic Activity*, it regulates economic and financial order, and even provides criminal liability for companies "without prejudice to the individual liability of the managing officers of a legal entity, establish the liability of the latter, subjecting it to punishments compatible with its nature, for acts performed against the economic and financial order and against the savings of the people" (Article 173, paragraph 5).

Criminal law has from the outset concerned itself with protecting the institutions of government and citizens' most basic interests. Over time, however, in addition to being relied upon to provide minimum standards of coexistence, it began to lend itself to the protection of new social and economic interests. A radical shift in government intervention strategies was enacted to combat the intimidating phenomenon

---

[10] In PEDRAZZI, Cesare. O Direito Penal das Sociedades e o Direito Penal Comum. *Revista Brasileira de Criminologia e Direito Penal*. Rio de Janeiro: Instituto de Criminologia do Estado da Guanabara, 1965, vol. 9, p. 133.

[11] Cf. OLIVEIRA, Eugênio Pacelli de, Coordinator. *Direito e processo penal na justiça federal: doutrina e jurisprudência*. São Paulo: Atlas, 2011, p. 69.

of organized crime that was perpetually working its will on politicians, journalists, judges, businesspersons, and others.

White-collar financial crime must be tackled alongside the image of the criminal and the social effects of that type of conduct. The conclusions advanced by Edwin Sutherland defined white-collar crime as a crime committed by an honorable person with social and professional prestige,[12] which may explain why this type of criminal conduct generates little in the way of social reaction.

Perhaps this phenomenon is due to the perception of minimal dangerousness of the criminal in the absence of direct violence or confrontation with the victim, or even because no physical harm is even contemplated. This brings us to the idea advanced by Thomas Lynch that serious crime is more ink-stained than bloodstained.[13] Hence, perhaps, serious crime involves a certain moral neutrality. According to Dr. José Ángel Garcia Brandariz, imprisoning financial criminals would not even result in the negative social stigma typically expected for those identified as criminals, given their personal and socioeconomic characteristics.[14]

Yet, it is by no means common knowledge that financial crime is more harmful to society than ordinary crime, given its penetration of our institutions and social organization. In truth, it ends up fostering ordinary criminality (corruption, unfair competition, fraud, etc.). This hampers enforcement efforts, due to widespread ignorance as to the harmful effects on society that result from the drain on already scarce financial resources.

Financial criminals do, in fact, have great potential for engendering other types of crime. They are highly adaptable within society and often enjoy considerable tolerance within the community, which leads to their increasingly daring and dangerous criminal behavior.

Furthermore, financial criminals actually run a sort of cost–benefit analysis on the gains to be had from unlawful conduct and possible sanctions imposed by the statutory system. By running the utilitarian calculation,[15] one could easily conclude that getting caught involves little or no consequence, given the tolerance and only formally rigorous nature of our criminal justice system. This technical sophistication is not found in ordinary criminals.

Claudia Santos Cruz argues that "theories of rational choice and situational prevention seem to fit like a glove. Their assessment of the costs and benefits associated with misconduct might dissuade them from engaging in it, should the opportunities decrease and the possibility of detection and punishment increase."[16]

---

[12] In CAVERO, Percy García. *Derecho Penal Económico—Parte General.* Lima, Peru: Grijley, 2nd ed., 2007, pp. 276–77.

[13] *Apud* José Ricardo Sanches Mir and Vicente Garrido Genovês (in *Delincuencia de cuella blanco.* Madrid: Instituto de Estudos de Política, 1987, p. 71).

[14] In *El delito de defraudación a la seguridad social.* Valencia: Tirand lo Blanch, 2000, pp. 80–81.

[15] FISCHER, Douglas. *Inovações no Direito Penal Econômico: contribuições criminológicas, político-criminais e dogmáticas.* Organizer: Artur de Brito Gueiros Souza. Brasília: Escola Superior do Ministério Público da União, 2011, p. 37.

[16] Cf. *O crime de colarinho branco* (da origem do conceito e sua relevância criminológica à questão da desigualdade na administração da justiça penal), 2001, p. 175.

The deciding factor is not will, but rather the impracticability of the behavior prohibited by law. No longer can we afford the luxury of complex theorizing over abstract hazards and social harm. Categories of financial crime have to do with increasingly complex regulatory situations and conduct that are legally intolerable, irrespective of the intentions of the criminal. Those intentions would only come out afterward, after the decision was made to break the rules of conduct binding upon us all.

Financial crime, given its scope and potential for damage, is consigned to the jurisdiction of the Federal Government—at least in countries such as Brazil and the USA, which rely on a dual justice system.

> A good portion of this jurisdiction is brought to bear upon crimes that are complex, some-times on account of the suspects or defendants involved—people of great economic or political power who, as a rule, operate within a network having international ramifica-tions—and other times because of the type of financial crime involved, be it corruption, influence-peddling, money laundering, etc. Its seriousness, its harm to society, and the threat it poses to institutions which safeguard the Rule of Law require a different balance between the rights of the accused and other procedural requirements of speedy trials and the duty of the State to prosecute and punish unlawful conduct.[17]

This harmful and unlawful behavior under federal jurisdiction requires recognition of financial criminal conduct as a violation of a negative legal duty, namely, that of refraining from illegitimately harming others or the public order, and of a positive legal duty, that one's behavior be conducive to the greater good of society. This progressive view of legality is increasingly accepted. It does, however, require more complex analysis, involving the said legal duties (both negative and positive) upon whose foundation a specific legal and criminal appraisal is constructed.

The reintegration of financial criminals into society must therefore center on making them rethink their behavior. If there is indeed any reasoning behind unlaw-ful conduct involving cost–benefit analyses of the outcomes to the offender, a given crime will be committed if and only if the expected penalty is outweighed by the advantages to be had from committing the act.[18]

Some rethinking is therefore in order on the application or restriction of mon-etary penalties. Inasmuch as their intimidating effect is quite low, they amount to little in the overhead of unlawful behavior. The exigencies of proportionality (grav-ity of the crime plus guilt) and the need for general deterrence require a response that is a better fit for serious financial crimes. Penalties should seek to elicit pre-ventive outcomes in both special and general cases because, if correct, timely, and inhibitory, they will be effective at the individual and societal levels, a point too obvious for society to ignore.

---

[17] OLIVEIRA, Eugênio Pacelli de. *Direito e processo penal na justiça federal: doutrina e juris-prudência.* São Paulo: Atlas, 2011, p. 71.

[18] For more on this, see Jesús-María Silva Sánchez (in *Eficiência e direito penal.* Coleção Estu-dos de Direito Penal. São Paulo: Manole, 2004, No. 11, p. 11) and Anabela Miranda Rodrigues (in *Contributo para a fundamentação de um discurso punitivo em matéria fiscal. Direito Penal Económico e Europeu: textos doutrinários.* Coimbra: Coimbra Ed., 1999, pp. 484–85).

Citing K. Polk, S.R.M. Mackenzie asserts that a fourth requirement ought to be added to the list to deter financial criminals. To the certainty, severity, and swiftness of punishment, one might add the calculation they make between the probability of punishment and the misconduct itself.[19]

In this context, it is worth remembering the notion stated by the United Nations Office on Drugs and Crime (UNODC) that, in order to help strengthen institutions to prevent crime and serve justice, it is essential to ballast the idea that no one can be above the law.[20]

3.3.1 Financial Action Task Force

Money laundering and terrorist financing trends and techniques evolve over time. The FATF carries out research on specific economic sectors and activities in order to determine their vulnerability to money laundering and terrorist financing. These studies, conducted by relevant experts from the public sector, are aimed at raising awareness and can lead to the development of new guidance on, or refinements to, the FATF Standards.

The FATF was created in December 1989 by the seven richest countries in the world, (G-7[21]), organized under the aegis of the Organization for Economic Cooperation and Development (OEDC). In June 2009, the FATF released an important report called *Money Laundering through the Football Sector,* a study of money laundering in sport that discussed several examples of areas that can be exploited by those who wish to invest illicit money through football.[22] Through the introduction of this report, the FATF reveals its growing concern over various legitimate sectors, such as the football sector, being subjected to the risk of corruption by illegal money.

Football in the past two decades has changed from a popular hobby to a global industry with growing economic importance. Thus, large sums of investments, specifically in football as compared to other sports, made the industry grow exponentially, and sadly, it also foreshadowed the sport's connection to organized crime. However, its regulatory framework has not been able to face the reality of crime within the sector. It is necessary to identify areas that can be used to launder money in order to draw the attention of governments, control agencies, and financial and

---

[19] In *Going, Going, Gone: Regulating the Market in Illicit Antiquities* In: http://www.mcdonald.cam.ac.uk/projects/iarc/culturewithoutcontext/issue18/gerstenblith_mackenzie_review.htm, accessed May 1, 2013.

[20] In: Promoting health, security and justice. 2010 UNODC Report, http://www.unodc.org, accessed April 13, 2010.

[21] USA, Japan, Germany, France, UK, Italy, and Canada, which have since been joined by Russia (G8).

[22] Cf. Money laundering through the football sector, available at http://www.fatf-gafi.org/media/fatf/documents/reports/ML%20through%20the%20Football%20Sector.pdf, accessed Feb 4, 2014.

nonfinancial sectors to the possible misuse of sport, and to gain a better understanding of the problem.

Football is by far the biggest sport in the world, practiced by over 265 million people. For example, the 2006 World Cup in Germany attracted more than a billion viewers, and the 2010 World Cup in South Africa broke audience records in a country with no footballing tradition—the USA. According to TV broadcaster ESPN, more than 6 million people watched the match between the USA and Algeria (the first phase), the highest for football in the history of that TV channel. There are records of an average increase of 68 % of the number of people in the audience during the first three games compared to the 2006 World Cup.[23] This global interest makes football attractive to criminals.

Thus, it is important to assess vulnerabilities in this context and, when in possession of such information, analyze the different types of misuses that allow the practice of money laundering. For the preparation of this study, the FATF and Committee of Experts on the Evaluation of Anti-Money Laundering Measures and the Financing of Terrorism (MONEYVAL) prepared a 2-day workshop in Monaco in November 2008, which was attended by Belgium, Brazil, Cyprus, France, Ireland, Italy, Monaco, Norway, Russia, Slovenia, South Africa, Switzerland, the Netherlands, the UK, the International Olympic Committee, and the Egmont Group. Additional countries were consulted and private sector groups (FIFA and the Union of European Football Associations—UEFA) also provided their cooperation and expertise.

A newspaper article and its quoted report depict the following fact: A player was being paid through a contract of 300,000 pounds that was not reported to the British IRS.[24] Contrary to what was reported, concealed payments were found that were made abroad to the player's agent, who then transferred the amount to the player himself, who was also abroad. It was discovered that the club was aware of the financial structure and avoided payment of 38,000 pounds to the British social security.

Another fact reported in both references revealed that a club paid a player a large sum of money for image rights that was placed in an offshore account located in a tax haven. In exchange, the athlete received shares of the offshore account. The British IRS realized that the club had not exploited the image rights and was even told by a consultant that nothing could be exploited commercially. It was surprising that the image of a top international player was considered to not have commercial value. The club maintained and renewed the image rights contract with the player. Later, he admitted that it was all an orchestration to enable tax evasion, and that the amounts were transferred to the tax haven as part of the player's income. The club agreed to pay, by way of taxation, 938,688 pounds, which demonstrated the

[23] Cf. Norte-americanos se rendem ao soccer e querem outra Copa. Folha de São Paulo, Copa 2010, D21, June 26, 2010; Os EUA aprenderam a gostar de futebol. O Estado de São Paulo, E7, June 27, 2010.

[24] In: Tax authorities provided information to football money laundering report. HM Revenue and Customs has confirmed it gave details of two tax evasion cases. The Guardian, July 2, 2009.

significant salary paid. The club also paid 404,480 pounds to use image rights in the coming years.

The discovery of these illicit financial transactions took place because of the cooperation between the UK authorities and FATF.

3.3.2 The FIUs, Law Enforcement Agencies, Securities and Exchange Commission, IRS, and Federal Reserve Banks

In Brazil, Bill No. 3443/2008, converted into Law No. 12.683 of July 9, 2012, which amended Law No. 9.613 of March 3, 1998, was hotly debated by many agencies that take part in the National Strategy for the Fight against Corruption and Money Laundering (ENCCLA). The ENCCLA is comprised of over 60 members, including many government agencies, such as Brazil's IRS (Receita Federal do Brasil), the Reserve Bank (Banco Central), the Ministry of Justice, state and federal attorneys' offices, the federal police, and state and federal courts. The ENCCLA strives to honor all international commitments entered into by Brazil, and keeps up with all countries that are members of the FATF.[25]

Among ENCCLA's recommendations is a need to close the loopholes that make money laundering feasible. Another recommendation is to require individuals in significant levels of trust (auditors, bank managers, insurance, real estate, capital goods brokers, etc.) to submit Suspicious Activity Reports to FIU, which are key to all crime fighting systems.

In closing with a set list of precedent crimes, ENCCLA attempted to fine-tune and update the law to the most modern standards of money-laundering legislation, thus providing preemptive asset forfeiture. Just as positive was the change requiring Suspicious Activity Reports from boards of trade, recordkeeping entities, and all those involved in mediating, brokering, or negotiating the trade of athletes (Article 9, XV). It was lax, however, in not including, for example, notification requirements on the part of sport clubs, sport federations, and sport confederations.

Law 9.613 of March 3, 1998, tried, as stated in its explanatory memorandum, to encompass the known phases of the practice of money laundering: placement,

---

[25] The most recent Argentine anti-money-laundering law (Law No. 26.683 of June 21, 2011), in addition to including self-laundering, increases the minimum sentence from 2 years to 3 years (while keeping the maximum at 10 years), requires the laundered money to have originated from a "criminal act" instead of a "crime," adds language to the Criminal Code making corporations subject to criminal liability, and establishes forfeiture of assets with no need for criminal conviction, provided illegal origin can be established, including cases of bankruptcy, flight, statutory limitations or the existence of any reason for suspending or terminating criminal proceedings, or when the defendant acknowledges the illegal source of the goods. (INFORME ANNUAL 2011. *Unidad de Información Financiera*. Buenos Aires: Departamento de Prensa, Ministerio de Justicia y Derechos Humanos/Presidencia de la Nación, 2012, pp. 24–26).

layering, and integration.[26] It also considered the fourth phase: cleaning one's tracks by closing bank accounts, withdrawing money, and simulating the sale of goods.

By the new language imparted by Law 12.683 of September 7, 2012, the crime of money laundering is now defined as:

Article 1 Concealing or disguising the nature, origin, location, disposition, movement or ownership of goods, securities or money derived directly or indirectly from the criminal offense.
Penalty: Three to ten years of imprisonment and a fine.
§ 1 The same penalty shall apply to anyone who, in order to conceal or disguise the use of goods, securities or money arising directly or indirectly from a criminal offense:
I—converts them into legal assets;
II—acquires, receives, trades, negotiates, gives or receives them in guarantee or in bailment, keeps them on deposit, negotiates or transfers them;
III—imports or exports goods at a price other than their true value.
§ 2 The same penalty shall also apply to anyone who:
I—makes use them in financial or business dealings of goods, securities or amounts they know or have reason to know are the proceeds of crime;
II—is a member of any group, association or office while aware that its primary or secondary activity involves the commission of the crimes provided herein.
§ 3 Such attempts are punishable pursuant to the sole paragraph of Article 14 of the Criminal Code.
§ 4 The penalty shall be increased by one-third to two-thirds if the crimes established in this law are committed in repeat offenses or through the criminal organization.
§ 5 The penalty may be reduced by one-third to two-thirds, and may be served under a work-release program or similar, or the judge may suspend the sentence or instead sentence the defendant to curtailment of rights if the first main second main accomplice or freely cooperates with the authorities, and provides information to assist in the investigation of the crimes, identifies perpetrators or identifies the whereabouts of the goods, securities or monetary proceeds of the crime.

Title 18 of the US Criminal Code establishes money laundering as a crime:

§ 1956 (Money Laundering)
(a)(1) Whoever, knowing the property that involved in a financial transaction represents the proceeds of some form of unlawful activity, conducts or attempts to conduct such a financial transaction which involves in fact the proceeds of specified unlawful activity—
(A)(i) with the intent to promote the carrying on of a specified unlawful activity, or with intent to engage in conduct constituting a violation of section 7201 or 7206 of the Internal Revenue Code of 1986; or knowing that the transaction is designed in whole or in part (i) to conceal or disguise the nature, the location, the source, the ownership, or the control of the proceeds of specified unlawful activity, or to avoid a transaction reporting requirement under State or Federal law, shall be sentenced to a fine of not more than $ 500,000 or twice the value of the property involved in the transaction, whichever is greater, or imprisonment for not more than twenty years, or both. (…) (2) Whoever transports, transmits, or transfers, or attempts to transport, transmit, or transfer a monetary instrument or funds from a place in the United States to or through a place outside the United States—(A) with the intent to promote the carrying on of specified unlawful activity, or (B) knowing that the monetary instrument or funds involved in the transportation, transmission, or transfer represent the proceeds of some form of unlawful activity and knowing that such transportation, transmission, or transfer is designed in whole or in part (i) to conceal or disguise the nature, the location, the source, the ownership, or the control of the proceeds of specified unlawful

---

[26] FATF adopted three phases.

activity, or (ii) to avoid a transaction reporting requirement under State or Federal law, shall be sentenced to a fine of not more than $ 500,000 or twice the value of the monetary instrument or funds involved in the transportation, transmission, or transfer, whichever is greater, or imprisonment for not more than twenty years, or both (…).
§ 1957 (Engaging in monetary transactions in property derived from specified unlawful activity)
Whoever, in any of the circumstances set forth in subsection (d), knowingly engages or attempts to engage in a monetary transaction in criminally derived property of a value greater than $ 10,000 and is derived from specified unlawful activity, shall be punished as provided in subsection (b).

Control agencies have an important role and responsibility in preventing the crime of money laundering. In this, there is usually the tender of a financial institution. There is a great concern with the records of these institutions, which highlights the role of Federal Reserve Banks. The banks themselves are obliged to report suspicious transactions to the FIUs (in the USA, the Financial Crimes Enforcement Network, or FinCEN; in Brazil, the Council for Financial Activities Control or COAF), and even to the SEC and the IRS in the USA or Comissão de Valores Mobiliários (CVM) in Brazil.

It is important to get adequate identification of customers in a way that it is not possible for them to realize that they are being subjected to an investigation by the financial institution. Philippe Broyer believes that "…the desire to conceal the true identity of individuals who are both the transferors and recipients of large washing operations makes them create numerous societies, which are often domiciled in offshore regulatory."[27]

The second phase of money laundering, layering, is the phase of control or concealment, aimed at distancing the origin of the money and breaking the chain of evidence. Criminals use international banks because all records of these institutions can be easily manipulated, thus further enabling their use by tax havens.

Offshore companies[28] are repeatedly used because of their insufficient or absent control, and even the creation of false leads and electronic wire international transfers. Offshore bank accounts make it possible to disguise the real controllers, since ownership is—according to the legislation in the countries in which they are located—evidenced by bearer paper, and partners or officers are simply proxies, often proxies for hundreds of companies of the same pattern. All of this amounts to creating a veil for the actual owners to hide behind. Their paper cannot be traded on the domestic market, nor cashed in without considerable expense and questions about possible complicity in money laundering directed at anyone who converts it.

The regulation of the Brazilian IRS (Instrução Normativa No. 568/SRF of August 9, 2005) requires legal entities domiciled abroad to enroll in the National Register

---

[27] Cf. BROYER, Philippe et al. La nouvelle économie criminelle: criminalité financière—comment le blanchiment de l'argent sale et le financement du terrorisme sont devenus une menace pour les entreprises et les marchés financiers. Paris: Éditions d'Organisation, 2002, p. 33.

[28] Citing TÔRRES, Heleno; Godoy, Arnaldo Sampaio de Moraes remembers that the "term offshore identifies those entities incorporated with capital owned by non-residents to hold positions on behalf of mother company outside the territory where it is headquarted" (In: Direito tributário comparado e tratados internacionais fiscais. FABRIS Sergio Antonio. Porto Alegre. p. 85, 2005).

of Legal Entities (CNPJ) whenever they have property and rights subject to registration in Brazil, or simply practice actions in the country, such as import financing, external leasing, letting, importing goods without hedging, currency loans granted to residents in the country, and investment in other operations (Article 11, part IV). It requires, therefore, a copy of the act of incorporation accompanied by a sworn translation authenticated by the Brazilian consulate domicile of the corporation. The Normative 748/RFB of June 28, 2007, also from the Brazilian IRS and regarding registration in CNPJ, in Article 15, objects to non-submission of the membership and officers for legal entities domiciled abroad, contrary to Brazilian companies, where complete identification is required. Thus, foreign companies can obtain the CNPJ and now operate in the country without identifying their partners and administrators, only indicating an attorney who often does not have any professional or commercial relationship with the company domiciled abroad. The only requirement is a document equivalent to the constitutive act, a mere statement issued by a public entity tax haven with the company name, date of opening, legal status, corporate status, and address. This is sufficient for the identification of directors and shareholders.

Bank regulation in the USA is highly fragmented compared with other countries in that most countries have only one bank regulator. In the USA, banking is regulated at both the federal and state levels. Depending on the type of charter the banking organization has and on its organizational structure, it may be subject to numerous federal and state banking regulations. Unlike Brazil, Japan, and the UK (where regulatory authority over banking, securities, and insurance industries is combined into one single financial service agency), the USA maintains separate securities, commodities, and insurance regulatory agencies from the bank regulatory agencies at the federal and state level. US banking regulation addresses privacy, disclosure, fraud prevention, anti-money laundering, antiterrorism, anti-lending, and the promotion of lending to lower-income populations. Some individual cities also enact their own financial regulation laws (for example, defining what constitutes usurious lending).

A bank's primary federal regulator could be the Federal Deposit Insurance Corporation, the Federal Reserve Board, or the Office of the Comptroller of the Currency. Within the Federal Reserve Board are 12 districts around 12 regional Federal Reserve Banks, each of which carries out the Federal Reserve Board's regulatory liabilities in its respective district. Credit unions are subject to most bank regulations and are supervised by the National Credit Union Administration. The Federal Financial Institutions Examination Council (FFIEC) establishes uniform principles, standards, and report forms for the other agencies.

Regulators of conditions for opening, maintaining, and operating deposit accounts should obtain, where possible, the identification of the controller, partner, and real beneficiaries who are domiciled abroad and wish to operate in the financial system.

The Brazilian Reserve Bank (Resolution No. 2.025/1993/BACEN), which establishes procedures for opening, maintaining, and operating deposit accounts, does not identify the real beneficiaries, but it does identify the partner or controller.

The integration phase in money laundering, i.e., the return of money to the place of origin, allows for the financing of illicit activity as well as various investments. In this respect, focusing on the so-called white-collar crimes, Maria Claudia Santos Cruz reveals that "a notable characteristic of organized crime is investing in legitimate activities, even for a cause as little noble as the need to wash the capital illegally obtained."[29]

Ricardo Liao, head of the Department for Combating the Illicit Financial Supervision Rates and International Capital Brazilian Central Bank (DECIC) reported that, at some point in the transaction, the money goes back to the banks. A representative of the Federal Bureau of Investigation (FBI), Carlos Alberto Costa, estimated that between US$ 500 billion and US$ 1.5 trillion is laundered in the world every year, not even including the drug trade and organized crime.[30] This represents 5 % of world production and more than a quarter of all international trade of goods.

In 2002 and 2003, a Study in the Federal Justice Council (CJF), chaired by the Minister Gilson Dipp of the Superior Court of Justice, with the participation of representatives of the Federal Court, the Federal Public Ministry, the Federal Police, and the Brazilian Federation of Banks (FEBRABAN), elaborated concrete recommendations to improve investigation and prosecution of the crime of money laundering through the cooperation of various sectors of the state and society responsible for the implementation of the law. This study is considered the forerunner of ENCLA (National Strategy Anti-Money Laundering and Asset Recovery), renamed ENC-CLA in 2007 (National Strategy for Combating Corruption and Money Laundering).

In turn, through Resolution 314 of May 12, 2003, the CJF created federal courts that specialize in crimes against the national financial system and money laundering. This was subsequently amended by Resolution 517 of June 30, 2006, to include crimes committed by criminal organizations requiring specialized competence by the courts. Delivering the recommendation of the Study Commission of the Council of the Federal Courts, the resolution recommended this specialization because the concentration of authorities with similar purposes increases the quality and speed of investigation and trials. Today, Brazil has several specialized criminal courts located in the main cities and towns of the country.

The FIUs are responsible for receiving, requesting, analyzing, and distributing to the appropriate authorities reports on financial information with respect to procedures presumed to be criminal under national legislation or regulations implemented to prevent washing. These units can be judicial, police, mixed (composed of judicial and police entities), or administrative.

Under the Ministry of Finance, the COAF was created in the wake of the FinCEN with the purpose of enforcing disciplinary action, applying administrative penalties, and receiving, examining, and identifying the suspicious activity reports (Article 14 of Law 9.613/98). It may request banking and financial information about people involved in suspicious activities (Law 10.701 of September 7, 2003, amending Law 9.613/98).

In the field of uncovering money laundering, it is relevant to require qualified and skilled labor that seeks to solve the puzzle behind this financial crime. Hence,

---

[29] Cf. SANTOS, Cláudia Maria Cruz. O crime de colarinho branco (da origem do conceito e sua relevância criminológica à questão da desigualdade na administração da justiça penal). Coimbra: Coimbra Ed., 2001, p. 89.

[30] Published article from Jornal Gazeta Mercantil, "Legislação", November 28, 2002.

it is important to have joint action by the bodies responsible for confronting this puzzle. The work for the prevention of money laundering in football must involve many agencies, such as those mentioned above, joining efforts to provide prosecutors elements that lead to effective investigation of the offense. Recognizing the need of institutions to work together in a coordinated manner and, if possible, with task forces, permits a quick response to the actions of organized crime, which historically have relied on the lack of organization within state agencies to facilitate their actions.

One can imagine the SEC having a minor role. However, the securities market, with the allocation of trading predominant shares of investment funds in equities, comprises a broad universe of securities created or issued by corporations and thus is included in the concept of security and properly registered with the competent agencies (e.g., SEC in the USA or CVM in Brazil). The main function of the securities market in the economy is to service the medium-term and long-term financing needs of companies, providing resources to able to fund expansion projects or advanced technology enterprises.

According to the SEC, its mission is as follows:

> to protect investors, maintain fair, orderly, and efficient markets, and facilitate capital formation. As more and more first-time investors turn to the markets to help secure their futures, pay for homes, and send children to college, our investor protection mission is more compelling than ever. As our nation's securities exchanges mature into global for-profit competitors, there is even greater need for sound market regulation.... The laws and rules that govern the securities industry in the United States derive from a simple and straightforward concept: all investors, whether large institutions or private individuals, should have access to certain basic facts about an investment prior to buying it, and so long as they hold it. To achieve this, the SEC requires public companies to disclose meaningful financial and other information to the public. This provides a common pool of knowledge for all investors to use to judge for themselves whether to buy, sell, or hold a particular security. Only through the steady flow of timely, comprehensive, and accurate information can people make sound investment decisions.... Crucial to the SEC's effectiveness in each of these areas is its enforcement authority. Each year the SEC brings hundreds of civil enforcement actions against companies and individuals for violation of the securities laws. Typical infractions include insider trading, accounting fraud, and providing false or misleading information about securities and the companies that issue them.[31]

The SEC and CVM have the particularly important functions of promoting the expansion and efficient operation of the stock market and stimulating permanent applications in the capital stock of public companies under the control of national private capital, while also protecting holders of securities and investors against irregular emissions and illegal acts of directors and shareholders of the companies. The CVM has legislative power by which it regulates the activity of various market actors. It also has punitive power, ensuring the full defense of the administrative procedure that may lead to the punishment of those who violate the rules issued by the agency or who practice fraudulent acts.

---

[31] Cf. Securities and Exchange Commission—SEC, http://www.sec.gov/about/whatwedo.shtml, accessed May 2, 2013.

Law No. 9.457 of May 5, 1997, amended provisions of Law No. 6.404 of December 15, 1976, which provides for the protection of corporations, and Law No. 6.385 of December 7, 1976, which provides for the protection of stock market securities and creates the SEC. Law No. 9.457/97 expanded the range of possible penalties and established the Commitment Agreement, which allows for the suspension of any procedure until the accused discontinues the tort and indemnifies any persons or entities harmed.

The SEC and CVM also have the responsibility of checking the accounting of public companies that control football teams. Public companies have to prepare their financial statements following the law and regulations. Thus, the expenses associated with the acquisition and training of football players should be well examined to ensure these companies are adopting best practices. Namely, the amounts paid by the pass and spending committees of businessmen must be properly classified and accounted for as an asset in the group assets (capital expenditure), and expenditures for training athletes must be classified as expenses for a specific period and should be recognized directly in the income of these companies.

The reassessment of intangible fixed assets (e.g., pass athletes) offends accounting principles. Should such reassessment occur, such as with loan agreements without payments that are entered into with related parties, a disclosed note should be adopted containing the minimum information needed for the performance of an unequivocal judgment about the fairness of such contracts. In the case of a football club shareholder offsetting receivables against future dividends, where uncertainties exist regarding its performance, the company has the duty to assess, on a periodic basis, the probability of loss to be recognized on that asset, and the necessary allowance due for doubtful accounts. The company should also properly check whether the accounting for intangible assets (e.g., "right to use the trademark") includes high values, since this would let the company know if the cost of such an asset is recorded correctly.

It is important to note that the Federal Accounting Council (CFC) has shown concern over the alleged poor quality of the financial statements of football clubs in Brazil.

There are investment funds created with the purpose of investing in sports-related activities, such as negotiating pass athletes exclusively for certain football clubs, by applying in shares of companies incorporated by these investments in sports ventures. It is important to detect whether these companies are sending monthly reports and half of these funds to shareholders, or if they are allocating expenditures of such funds to shareholders differently than what was recorded. Through the issuance of shares, it is possible to acquire investments from several individuals for investment in portfolios of assets available in the financial market.

It is important for the fund administrators to perform their duties because they are responsible for managing the funds and providing information about shareholders, who must be properly identified to SEC or CVM. The portfolio manager has the same function as the fund administrators, if third parties perform this management function. The main duties of the administrators include: (1) regularly disseminating information to investors, in frequency, in time and with content defined, (2) advertising the share value and shareholders' equity of the opened fund on a daily basis, (3) remitting monthly account statements containing the fund's name, registration number, and even the name, address, and registration number of the administrator, (4) sending information about funds, which will be available for Internet consultation to SEC or CVM, (5) annually sending the financial statements, together with the report of the independent auditor, within 90 days from the end of the year to which they refer, and (6) immediately communicating any act that occurred or fact related to the fund operation or the value of its portfolio.

In turn, the IRS has the responsibility of verifying that the clubs report revenues from their economic, financial, and business activities, such as the provisions of services and/or sale of goods, even if only to members; renting of halls, auditoriums, sports courts, swimming pools, sports fields, outbuildings, and facilities; ticket sales; and amounts received as investments.

It should be noted that the immunity granted to sports bodies, incorporated as nonprofit associations, is not a subjective immunity associated with the activity of the entity, but objective, linked to a specific operation on the activities themselves. The revenue earned, even if recorded in the stadium, restaurants, and cafeterias, should only be considered marginal if necessary to maintain sport clubs, being outside the scope of tax liability. For example, entities with recreational and cultural aspects beyond sport often are supported by tax exemptions, such that the revenues derived from these non-club activities should be subject to tax incidence. Thus, revenues from parking and food sales have an exchange character, which prevents tax. Meanwhile, revenues from activities related to sport, with a clear professional character, obviously are not social services and cannot fit within the tax exemption, because clubs cannot be characterized as nonprofit organizations.

In turn, the Federal Reserve Bank may exercise substantial control over the operations and rights of athletes in foreign clubs, since it has to check whether conditions in contracts between international clubs abroad and domestic ones are based on foreign money exchange transactions. Financial institutions involved in the flow of money from or to international destinations should refuse transactions if there is negative information about the owners and origin of the money.

3.3.3 Fédération Internationale de Football Association, National Leagues, and Clubs

Football is administered by FIFA, which is based in Zurich, Switzerland. It is a private entity, governed by Swiss law, controlling the whole world of football practice through a confederation system. It has the authority to promote and develop football for the world. Each country has an associate that must submit to FIFA's rules and laws.[32] Moreover, FIFA has a clear responsibility to safeguard the reputation and integrity of the sport sector. Thus, one of the objectives of FIFA is to prevent practices that may jeopardize the integrity of any sport.

[32] Cf. Estatuto da FIFA, http://www.fifa.com, accessed May 5, 2013. Article 2 contemplates the FIFA objectives.

For this reason, FIFA approved the Code of Ethics in 2004 (later revised in 2006), which enabled the creation of the new Ethics Committee as an important member agency. As part of its efforts to reinforce ethics in sport, FIFA offers technical support through the company Early Warning Systems GmbH, founded specifically to monitor sport betting and to prevent negative effects of unethical behavior in the football games. An example of an alleged ethical violation was when Brazilian referee Rodrigo Braghetto, who was scheduled to referee the final of the São Paulo Championship between Santos and Corinthians on May 19, 2013, was cut on May 17th for being a member of a company that provides services to Corinthians. The removal occurred by Paulista Football Federation (FPF) only after the release of this information in the "Blog Paulinho" (from Paulo César de Andrade Padro).[33]

It is important to set up a supervisory body to monitor the football sector closely, including its management of clubs that often bear debts incompatible with effective financial capacity. FIFA is the guardian of such supervision and includes six confederations: Asian Football Confederation in Asia and Australia (AFC), Confédération Africaine de Football (CAF), Confederation of North, Central American and Caribbean Association Football (CONCACAF), Confederation Sudamericana de Fútbol (CONMEBOL), Oceania Football Confederation (OFC), and Union of European Football Associations (UEFA).[34] UEFA is by far the largest of the six continental confederations.

However, many criticisms have been leveled at FIFA. For example, Andrew Jennings, investigative Scottish reporter and author of *Foul! The Secret World of FIFA*, published in 2006, reported that senior FIFA leaders received bribes. This was recognized by the Swiss Court, but its sentencing only entailed the payment of legal costs and the preservation of the parties' identities. In his new book *A Comparison between FIFA and Organized Crime* that is yet to be published, the author reveals that the money paid by FIFA per year for each country to invest in football (US\$ 250,000) is never audited, that corruption is indeed recognized, that countries do nothing to curb the corruption due to the fear of being suspended, and, finally, that the internal audit committee contains members who have already been investigated.[35]

FIFA's instant reaction to the mere announcement of summons on June 30, 2010, from the French National Assembly (Palais Bourbon) provoked astonishment. The Committee on Cultural Affairs called the former national coach of France, Raymond Domenech, and the former president of the French Football Federation, Jean-Pierre Escalettes, to give explanations on the performance of the selection in the World Cup in South Africa. FIFA's president declared that the French team could

---

[33] In Ligação com o Corinthians faz FPF tirar juiz da final. Paulista. Empresa de Rodrigo Braguetto tem contrato com o clube há dois anos. *Folha de São Paulo*, D2 esporte, May 18, 2013.

[34] Founded in June 1954, UEFA is an umbrella organization representing 53 national association members. It is the parent body of European football and is one of six confederations under FIFA.

[35] A ginga perfeita dos donos da bola. A FIFA controla o dinheiro, marca os adversários e dribla a Justiça. Entrevista. *O Estado de São Paulo*, J4, *aliás*, June 6, 2010.

be suspended.[36] Congressman Renaud Muselier defended the parliamentary debate by stating that "winning or losing is part of sports life, but from the moment that there are consequences in terms of national and international image, our job is to try to clarify."[37]

An attempt at corruption involving Australia being chosen as the country to host the 2022 World Cup was unveiled by the Australian newspaper *The Age*. The article said that gifts, trips, and projects were purchased for committee members in charge of the FIFA World Cup site selection. The Federation of Australian Football (FFA) spent $ 50,000 on 24 necklaces that were given to 24 female members of the site selection committee, a fact admitted to by its chief executive, Well Buckley. However, he argued that it is a "common practice among governments, business and sports organization to give gifts," and that it would follow FIFA's rules.[38]

FIFA's attempts to obtain vital information through the Transfer Matching System (TMS)[39] are valid, but not enough. It is an important tool for obtaining information on the international transfer of players, which was previously restricted to business stakeholders. Through this system, over 30 pieces of information are recorded online, such as player history, clubs involved in the business, payments, values, contracts, and other kinds of information.[40] For instance, one can check to see whether the contracted amounts were allocated directly to the involved parties before such amounts have been reported to the recipient bank accounts. This is very important, especially when the information is electronically available.

According to FIFA, TMS aims to:

1. Enable clubs to confirm the terms and conditions of player transfers
2. Facilitate the transfer of player registrations between Associations
3. Help safeguard the protection of minors
4. Provide information and decision-making tools to key stakeholders
5. Train and support key stakeholders
6. Monitor player transfer activities and investigate alleged breaches of transfer regulations
7. Enforce adherence to the transfer regulations through a specific sanction system, presenting breaches of those transfer regulations and proposing sanctions to the competent FIFA bodies

---

[36] In: Audiência na França faz Fifa ameaçar país. Ex-técnico e ex-presidente da federação francesa vão à Assembleia Nacional hoje dar explicações sobre fiasco. Folha de São Paulo, Copa 2010, D18, June 30, 2010.

[37] Cf. Domenech culpa imprensa por crise. Segundo deputado, treinador francês perdeu controle da equipe após jornal pôr na capa palavrões de Anelka. Folha de São Paulo, Copa 2010, D21, July 1, 2010.

[38] AUSTRÁLIA é acusada de subornar a Fifa para receber a Copa-2022. Federação deu colares de pérolas a comitê que escolherá a sede. *Folha de São Paulo*. Copa 2010, D13, July 1, 2010.

[39] In FIFA Regulations on the Status and Transfer of Players, http://www.fifa.com/mm/document/affederation/administration/01/95/83/85//regulationsstatusandtransfer_e.pdf, accessed June 21, 2013.

[40] Cf. FIFA, Transfer Matching System. In http://www.fifa.com/aboutfifa/organisation/football-governance/transfermatchingsystem.html, accessed June 20, 2013.

However, FIFA should not be the only recipient of such data, since the autonomy guaranteed to sport, and especially conferred by the Brazilian Constitution, limits it to the sport's organization and operation. It is essential to create certain obligations, like reporting of suspicious transactions to the FIUs, upon clubs, federations, and confederations, and upon those who provide advisory, auditing, bookkeeping, and consulting in this area. There are records of money laundering occurring within clubs during negotiations of international money transfers in various countries. In fact, according to the FATF, clubs are deliberately being used to launder money. That is why it is insufficient to create a system where information is only supplied by these clubs without any control over them. Authorities should not be replaced by FIFA, an organization that sometimes acts purely with commercial and private interests. FIFA data are not public and not easy to obtain, and authorities would be forced to request international legal cooperation to access them because FIFA is headquartered abroad.

The national associations have a responsibility to discipline, coordinate, and administer football in their respective countries. Such entities at the national level are considered to be the first regulators in the country, but they must still comply with specific regulations established by FIFA. In turn, the national associations can be subdivided into regional bodies. Clubs are considered cells that are at the base of each regional body. The oldest national association is the English Football Association (EFA), founded in 1877. Throughout the history of FIFA, its statutes have been submitted to several reviews, which have allowed the statutes to have the mark of modern law and transform into an increasingly comprehensive body of work. They determine the basic laws of world football, including numerous rules about competitions, transfers, illegal drug use, and a variety of other subjects. These bylaws were approved at the 59th FIFA Congress in Nassau, Bahamas, on June 3, 2009, and became effective on August 2 of the same year. Changes in the FIFA Statutes can only be made by a Congressional session and require a three-fourths majority of national federations present and entitled to vote. This makes FIFA Statutes and their implementing regulations equivalent to a constitution of the governing body of international football.

The Brazilian Federal Constitution established the autonomy of sports bodies, along with the mandatory exhaustion of remedies through the sports bodies with respect to actions regarding discipline and sporting competitions before resorting to the courts, as well as fixing a period of 60 days from the filing proceedings before a final ruling is made by the Sport Court (Article 217). Sport Court has an administrative nature and does not belong to the judiciary branch, thus constituting an independent and autonomous unit. It seeks to expedite timely decisions in cases involving violations of competition rules and disciplinary offenses.

In Brazil, the Pass Law, Law No. 6.354/76, later revoked, provided for labor relations for professional athletes in football, setting the "pass" as "the sums due from one club to another club for letting the athlete go during the term of the contract or after its completion, in compliance with relevant sport rules" (Article 11). The sport bond remained even after the termination of the employment contract, causing the athlete to be stuck with the club, and was justified by the need to reward the investment in staff training and professional players. The "pass," under the law, would end with the termination of the contract once the athlete reached the age of 32 and had worked for his last employer for 10 years. The "pass" holder would not necessarily be a club, but any person or entity who invests in the training of sporting talents. The freedom of the worker eventually prevents the right to the "pass," since this forbids the exercise of activities in other associations, except with the liberality of its owner.

Law 8.672 of July 6, 1993, called the Zico Law, was intended to allow clubs the chance to become companies. In turn, Law 9.615 of March 3, 1998, called the Pelé Law, replaced the Zico Law and was later amended by Law 9.981 of July 14, 2000, called the Maguito Vilela Law, which governs the conclusion of partnerships among clubs, sponsors, and investors. Article 2 of the Maguito Vilela Law revoked the chapter of the Pelé Law (Articles 59–81) devoted to "Bingo," a specific type of gambling similar to casinos.

Law 10.671 of May 15, 2003, added financial transparency of management, established offenses, and considered sports as a cultural expression of the country. Law 10.671 basically includes the following: the need for obedience by constitutional duty and express legal provision of national and international standards (Article 1, § 1), sovereignty in the organization of sport, national identity and transparent financial management, ethics, and social responsibility (Article 2—I, II, and sole paragraph), the obligation of the National Sports System to be established autonomously, based on freedom of association, requiring quality standards and integrating the Brazilian cultural heritage, which is considered a high social interest (Article 4—IV, §§ 1–3), and the incidental nature of the link between sport and its athletes, dissolving it to all its legal effects, valid beginning March 26, 2001 (Article 28, § 2, and Article 93). Finally, it establishes the prior right for the first renewal of the original professional contract, with a maximum term of 2 years by the athlete's training entity (Article 29, heading).

Law 12.385 of March 16, 2011, altered the Pelé Law, and clarified some points about professional sport practices.

Law 12.868 of October 15, 2013, added an article to Law 9.615 (Article 18-A) establishing that nonprofit clubs and social entities can get public funding only if they exhibit real transparent management and permit access to their accountability mechanisms (except for confidential contracts).

According to Álvaro de Melo Filho, the obligation to obey international standards stems not only from legal necessity but also from voluntary membership by confederations and federations in FIFA.[41] FIFA rules are ultimately imposed on several countries and do not allow any discussion, lest there be a violation of sovereignty. This opinion is also supported by Philip Legrazie Ezabella.[42] However, it should be understood that FIFA's standards in each country only apply to the organization and functioning of sports bodies, due to the autonomy granted to these entities.

Ricardo Georges Affonso Miguel considers the possibility of a player filing complaints with the courts for authorization to transfer from one club to another and obtaining registration by FIFA and the federation of each country outside the period

---

[41] Cf. MELO FILHO, Álvaro. O Novo Direito Esportivo. Brasília, São Paulo: Cultural Paulista, 2002. p. 69–70.

[42] In: Agente FIFA e o Direito Civil Brasileiro. São Paulo: Quartier Latin, 2010, p. 56–57.

of registration for athlete transfers.[43] The author demonstrates with his conclusion that these are hypotheses that must prevail over the principles and rules of a country, although whether they would prevail over FIFA's rules is a different discussion.

Due to national sovereignty, only the law, from which one can extrapolate the above ceiling of organization and operation, can contemplate obligations that would subject someone to penalty if not completed. FIFA, a private entity, in fact deals with public interests and is considered to be under the law's jurisdiction. On the one hand, questions arise as to the legitimacy of specific FIFA regulations in the country and their scope to better serve the sport (i.e., its organization and operation). On the other hand, FIFA regulations can also be interpreted as only subject to specific legal rules, such as the obligation to license FIFA agents only to individuals, the prohibition of clubs and coaches to use services without a license, and the obligation to ensure operations (see FIFA Players' Agents Regulations).

Article 26 of the Pelé Law provides for freedom of organization of professional activity by athletes and entities, while Article 27 says that, in order to obtain public funds, entities administering sport should perform all acts necessary to enable exact identification of the financial situation, provide rescue plans and investments, ensure the independence of the boards of directors and supervisory entities, adopt professional and transparent models, and develop and publish its financial statements, after they have been audited by independent auditors. It also determines the accountability of leaders when deviating from established goals, i.e., acting for themselves or others, thus foreshadowing the creation of clubs as business enterprises.

It is interesting to note that the Pelé Law determines that sports bodies will be qualified as business companies, even when they do not qualify as a business, pursuant to Article 990 of the Brazilian Civil Code, which would require the partners to respond jointly and severally to legal obligations. Thus, members began to be held jointly liable for the debts of the social clubs. Article 27 stipulates the prohibition to the same person or entity to control a single team. With respect to the sport bond, as already said, the Pelé Law establishes ancillary to their employment bond, dissolving the bond when necessary for all legal purposes (Article 28, § 2).

Article 28, § 7 addresses the issue of sports agents, prohibiting the granting of proxies to them with a term exceeding 1 year. However, the automatic renewal of an attorney in case of silence was not sealed. Such an article has not extended the fence to the agent contracts.

In this regard, Álvaro de Melo Filho says:

> The device would be more adjusted to the reality of professional sport in Brazil, if this prohibition was extended to "contracts for services," which are the legal instruments more effectively used by agents and managers to link and "hold" the athletes with potential market valuation of its professional sports. If too much, it is suggested to extend two (2) years

[43] Cf. A possibilidade de contratação do atleta menor de futebol e a utilização do instituto da antecipação de tutela para transferência do atleta de futebol. *Revista do Tribunal Regional do Trabalho da 1ª Região*, Rio de Janeiro, v.21, n.47, p. 103–116, jan./jun. 2010.

or restrict temporal delimitation, to coincide with the deadline set by FIFA, in art. 12, paragraph 2 of the Rules of Players' Agents.[44]

Article 29 of the Pelé Law states that that an entity who wants to form a business relationship with an athlete has the right to sign an initial professional contract with the athlete that includes a term not exceeding 5 years, with the option to renew the contract with the same conditions. The contract may last for a period of 2 years or, based on a systematic interpretation of Article 30, for 5 years. Article 29 also includes the right to recover the costs of training athletes (Article 29, §§ 5 and 6). Article 31 deals with the termination of employment of a professional athlete when there is delay in payment of wages, in whole or in part, for a period not exceeding 3 months. The athlete is free to transfer and request a termination fine. Article 34 establishes the duties of the national administration that registers the contract of employment between the sports entity and the athlete. With regard to transfers of professional athletes, Articles 38 through 40 determine that a transfer is valid if there is formal, written consent.

With respect to broadcast rights, the Pelé Law states that it is up to sports entities to authorize and/or prohibit such broadcasting, as well as to determine the price for distribution if the broadcasting is allowed. Since fans are treated like consumers, the law also brings focus to the Consumer Code (Article 42). The law prohibits professional sports (amateur is allowed) in schools, colleges, universities, the armed forces (Army, Navy, Air Force), and for children younger than 16 years old (Article 44). It is also necessary to have insurance contracts for accidents at work (Article 45). Financial statements and independent audits are required from sport leagues and entities administering sports, and those statements and audits must be submitted to the National Sports Council (CNE; Article 46-A).

The Pelé Law also addresses the Sports Court, an administrative entity that can rule only about disciplinary violations and sporting tournaments or competitions (Articles 46–55). The Sports Court does not decide every case related to the sports sector. Thus, the analysis of criminal matters, civil matters, labor law, tax law, and FIFA's agents is not delegated to the Sports Court, unless the matter incorporates an issue about discipline and/or athletic competition. For example, cases regarding corruption or misrepresentation in making contracts with agents could fall within the purview of the Sports Court. It should be noted that, in the rush to discipline football, private entities involved in matters unrelated to sport have tried to use Sports Court as a way to defend their commercial interests.

Under the Pelé Law, leaders and sports administration entities do not exercise publicly delegated functions (Article 82). The law also establishes that the international sporting entities must receive the same treatment as national sporting entities (Article 83).

Law 9.981 of July 14, 2000, amended the Pelé Law, stating that the CNE is a collective body for regulation, deliberation, and advice, directly linked to the office of the Minister of Sport, which is composed of the Minister of Sport, the Secretary

---

[44] Felipe Legrazie Ezabella, in: Agente FIFA e o Direito Civil Brasileiro. São Paulo: Quartier Latin, pp. 58–59.

of the Secretariat of the National Sport, a representative from sport administration entities, two representatives from the sports profession, a representative for athletes, a representative from the Brazilian Olympic Committee (COB), a representative from the Brazilian Paralympic Committee (CPOB), four representatives for educational sports appointed by the President of the Republic, a representative from the sport state secretaries, and three representatives appointed by Congress. They have a 2-year mandate term that can be renewed (Articles 11 and 12-A). In Brazil, the CNE ensures the implementation of legal and regulatory provisions. It gives opinions and recommendations and proposes measures for the application of public funds from the Sports Department (Ministério dos Esportes).

Law No. 10.671 of May 15, 2003, called the Fan Defense Statute, brought important innovations, notably with respect to administrative and financial transparency in economic activity within the sport sector. Fans have the right to an organized tournament, safe locations for competitions, safe transport, and good quality food in stadiums. The Fan Defense Statute makes clear that the sports organization is founded on freedom of association and integrates the Brazilian cultural heritage with high social interest. The federal courts have jurisdiction to decide about it (Article 4). The law established the right to transparency and publicity in the organization of sport through administrative bodies linked to professional football (Article 5). There is mandatory disclosure of income and the number of fans in stadiums (Article 7). By stipulating the guidelines of Sports Court, it stated that this court has to be guided by impartiality, speed, publicity, and independence principles. Proceedings in secret are not allowed, under penalty of nullity or dismissal of the case (Articles 34–36). As modified by Law No. 12.299 of July 27, 2010, there was a concern to inhibit violence in stadiums and in their vicinity. For that, for example, Article 2-A, sole paragraph, establishes the duty to complete identification of all members or associates of organized supporters (full name, photograph, affiliation, civil registry number, record number in the IRS, date of birth, marital status, occupation, address, and education). They can be sued for objective liability due to damages occasionally caused (Article 39-B). Finally, Law 11.438, of December 29, 2006, sets limits on individuals and corporations for tax deduction (Article 1, § 1, I and II, 1 % the tax for companies and 6 % for individuals).

FIFA, with its scope of authority to maintain a good atmosphere in football, initiated the creation of Special Courts for short-term trials and decisions. According to Judge Cherril Loots, who was responsible for a Special Court in the World Cup held in South Africa in 2010, two people who stole a photographer's equipment were sentenced to 15 years in prison. A man who stole a cell phone from a foreign tourist was sentenced to 5 years in prison. Special squads gathered evidence for prosecution in record time, based on legal provisions determined by FIFA. South African authorities reportedly arrested two Dutch women who wore miniskirts with the colors of a European brewery different from the brewery that sponsored FIFA. They were ultimately released immediately after an out-of-court agreement was made.[45]

---

[45] FIFA exigiu mudanças na justiça da África do Sul. Jornal da Globo, TV Globo, June 23, 2010, http://g1.globo.com/jornal-da-globo/.

It is no different in Brazil. Even though some points are of dubious constitutionality, the Brazilian government, on behalf of state and local authorities, applied to host the 2014 World Cup. Through its offering of guarantees to FIFA in May and June 2007, valid until December 31, 2014, Brazil assured FIFA of the following:

a. Issuance of unconditional and unrestricted entry visas and output to a list of people, including members of FIFA's delegation and all allied business staff, local broadcast teams, agencies with broadcast rights, staff from FIFA's trade partners, providers of accommodation, FIFA's ticketing partners, and FIFA's IT solutions.
b. Temporary import and subsequent export, without restriction, of any and all goods necessary for the organization, ensuring quick passage and no charging of any customs duties, value-added tax, or other charges or government taxes (federal, state, or local), including exemptions for FIFA's licensees and their staff, FIFA's business partners and their staff, FIFA's official partners of hosting services, and representatives of the media.
c. Prohibition on collecting any taxes or fees from FIFA, its subsidiaries, the teams, the referees, FIFA confederations, local broadcasters, agencies with broadcast rights, FIFA's business partners, FIFA's official partners of hosting services, FIFA's ticketing partners, and FIFA's IT solutions.
d. Unrestricted exchange of foreign currencies to and from Brazil, with guarantees of full import and export to FIFA members, business affiliates, transmission teams, hosting services, media representatives, and spectators.
e. Priority treatment for immigration, customs, and check-in to all members of FIFA's delegation, its managers, and teams.
f. Prohibition of "ambush marketing," a strategy that consists of taking advantage by invading an event or its advertising space without a supporting contract with those in charge of the event who hold advertising rights, whether or not the marketing qualifies as an intrusion. However, the implementation and execution of an exclusive area for advertising, such as street trading in a 2 km radius around each location, including the above airspace, are acceptable, as long as FIFA and its designees can reserve an exclusive right of exploitation.
g. Changing of names of stadiums at the discretion of FIFA.
h. Granting of special powers to authorities who have no judicial power, such as searches, seizures, forfeitures, arrests, and destruction of property, in order to enforce prohibitions pertaining to the exclusive shopping areas.
i. Unrestricted granting of all media rights, marketing rights, trademarks, and other intellectual property rights exclusively to FIFA.
j. Imposition of civil liability solely on the Federal Government, including fees for any damages, litigation, or claim costs brought by third parties against FIFA, its officers, employees, and consultants.

The World Cup General Law, Law No. 12.663 of June 5, 2012, contemplated the trade restrictions mentioned above, the visas granted without restrictions to FIFA representatives and their guests, and the responsibility imposed on the Federal Government for any damage that occurs during the World Cup. Furthermore, it established that the misuse of official symbols owned by FIFA as well as any intrusions caused by ambush marketing should be treated as crimes. It dismissed public procurements for telecommunication services, but considered the overlap of national holidays with the days when the Brazilian team played games. For example, it reallocated school holidays to the time between the opening and closing of the 2014 World Cup.

Federation rights involve the admission and registration of players and, more particularly, the granting of the player's pass, which establishes the right of the player to compete in the competition. The person who owns this pass or these rights actually owns a player's transfer rights. It is up to the National Leagues to gather this information from federations and sport clubs. In Brazil, this function belongs to the Brazilian Football Confederation (CBF). The CBF is an educational nonprofit organization based in Rio de Janeiro that embraces all Brazilian football associations and federations. This is one of the most powerful organizations in the country, always giving the final word in a league.

The CBF, formerly called Brazilian Sports Confederation (CBD), is a private entity founded on September 24, 1979. Its accounts are kept secret, despite the popularity of football in Brazil. The CBF is responsible for organizing nationwide tournaments, such as the Brazilian Championship of series A, B, C, and D, and the Brazil Cup. This entity has the right to dispose of any gains arising from the Brazilian team, which it manages. The CBF does not have any obligation to disclose, for example, the financial amount of sponsorship, much less the amounts received from media under transmission rights and rights resulting from friendly games. However, the CBF must also shoulder the costs of salaries of permanent staff and all the players while they are at the disposal of the national team.

For many years, the president of the CBF, who spoke and acted on behalf of CBF and determined the course of its actions, remained the same. All professional football teams are affiliated with the federation of each state. Since the federations have the power to elect the CBF president, and votes from the teams in a federation carry the same weight, even if in the second division, great political ability is required to get the votes of smaller or weaker clubs, regardless of any real popularity and prestige.

Discussions and protests have occurred regarding the CBF's power, such as the one that occurred in 1987 when 13 major Brazilian clubs decided to create a sort of parallel league not recognized by CBF due to disagreements regarding the division of television revenues. However, any inappropriate conduct committed by the CBF has minimal relation to suspicious activities involving football. CBF's performance is limited to the issuance of Certificates of International Transfer of Players, where it may ignore conditions and sums of transactions. The entity does not interfere in the athletes' transactions between clubs. Like the federations, the CBF only receives standard contracts between the club and the athlete. For these reasons, the CBF and/or federations have no way of informing others about payment loans, sums, and contractors.

Most clubs have no risk-assessment strategies concerning organized crime. The ideology that "players are the club's main assets" is not a sophisticated way of recognizing and gaining control over organized crime.

Law 10.672 of May 15, 2003, amending the Pelé Law, obliges football teams to publish their accountability. Today, it is possible to obtain accounting and management indicators of main associations of the country. Therefore, there is greater transparency and tax responsibility on the part of leaders. Resolution No. 1.005/04 of the CFC establishes the form that the accounting and financial statements due to sporting entities should take.

With the end of the "Pass," concepts of federal rights and economic rights have come to the forefront. Federal rights refer to rights exclusive to private clubs, while economic rights can be exercised by both clubs and entrepreneurs, but their content and purpose are similar to federal rights.[46] These concepts have led to some confusion in contracting. Moreover, historically one main source of income for the clubs has been the transfer of federal rights, and, given their subjectivity of value, the transfer of federal rights can be manipulated for criminal purposes, particularly money laundering.

According to Casual Independent Auditors, the revenue of the 20 most profitable Brazilian football clubs in 2008 reached the milestone of more than 1.4 billion reals (Brazilian currency), a 35% increase from 2007. Of this total, between 60 and 70% of the profits came from the sale of federal rights, whereas the traditional source of revenue, ticket sales, composed less than 10% of the total revenue.

Complementing this increase in revenue is the blatant inconsistency between the numbers reported by various sources that strongly indicates the presence of money laundering. For example, the Report of International Transfers of the Brazilian Football Confederation in 2007 indicates transfers abroad of 1,085 football players, but the Brazilian Federal Reserve's records (Banco Central) indicate the completion of 546 transactions in 2007. This proves that the payment for many of these shipments abroad involve payments in cash or other clandestine forms. In turn, the costs claimed by these entities have been higher than their revenues, despite the high revenue amounts involved, which demonstrates a lack of planning and professional administration within the sport sector.

3.4   France: Control Model of Accounts in Football

One interesting experience worth reflecting on comes from France, where the Direction Nationale du Contrôle de Gestion (DNCG) was created with the purpose of professionally handling the financial management of clubs.

The DNCG is composed of members of the French Football Federation (FFF), the Professional Football League (LFP), the Union of Professional Football Clubs (UCPF), the National Union of Professional Footballers (UNFP), the National Union of Educators Domains and Technical Football (UNE.CATEF) and the National Union of Administrative Sector and Assimilated Football (SNAAF). This ini-

---

[46] Cf. FILHO, Álvaro Melho, apud Felipe Legrazie Ezabella (In: Agente FIFA e o Direito Civil Brasileiro. São Paulo: Quartier Latin, 2010, pp. 49–50).

tiative allows for an unbiased review of the appropriate business performed within the clubs, checking each club's investor solvency and the source of the involved funds. Comprised of a body of accountants concerned about the solvency of clubs and investors, it oversees the origin of these resources and collateral warranty. The DNCG has demanded personal and real guarantees from investors, requiring the submission of contracts and accounts of clubs for its approval.

Professional clubs are subject, in each period, to the control of their legal and financial situation by the Control Committee of Professional Clubs, one of the three committees that integrate the DNCG, besides the Federal Control Club and the Appeals Commission. Control is exercised independently within the mission entrusted to the DNCG by the Law of July 16, 1984. The DNCG, exerting this control over professional clubs, has the main objective of ensuring the stability and fairness of competitions, in particular, by verifying that investments of each sports club do not exceed their financial capabilities. This assessment is made on the basis of financial, historical, and forecasted data reported by clubs.

The DNCG prioritizes the accounting rules, which are aimed at improving the image of true accounts of clubs. The Board wishes to establish a body of rules that ensures proper comparison of the accounts between the clubs and provides a higher level of confidence to investors. Thus, the Control Committee of Professional Clubs performs its mission with judicious application of the texts governing its action and strong concern for maintaining the impartiality of their decisions.

The DNCG does not actually audit the accounts of clubs, but has the primary mission of controlling the legal and financial situation of the clubs. Therefore, a concern exists regarding the integrity of the information provided to it. If the DNCG detects fraud or error in the accounting, it is forwarded to the relevant bodies. However, in cases where provisions on accounting issues are violated, the DNCG can, according to the severity of violations, do any of the following: (1) fine and prohibit the club's participation in the national championship, (2) determine the loss of championship points, (3) disapprove contracts through the prohibition of hiring players, and (4) check the controlled recruitment in comparison with the budget constraint in the face of the possibility that payroll has occurred before the player has signed or will sign contracts. These contracts must be approved or ratified by the DNCG, and failure to do so leads to the suspension of the leaders and the demotion of the clubs to a lower division.

This oversight by the DNCG is required due to the risks related to transfer transactions. The DNCG is able to check the papers of internal controls of clubs, and their effectiveness, in order to identify best practices inside them. The DNCG can therefore do the following: (1) require all relevant information from clubs for auditing the consolidated accounts and check compliance with legal provisions, (2) propose changes in terms of accounting levels, (3) approve or endorse the various aspects of controlled recruitment, including payroll and making site inspections, (4) examine financial situations, (5) enforce sanctions, (6) require clubs to provide accurate information referring to relevant economic events that are likely to adversely affect their financial position, and (7) perform independent audits. This oversight ultimately improves the financial credibility of the clubs.

Football should always require transparency and should adopt this professionalism and transparency during trading activity. The engagement of financial transparency of professional football clubs is materialized through the publication of the accounts of clubs attached to the DNCG report. The publication of the accounting allows for the gathering of financial information that can be examined by DNCG on all professional clubs from the same country. There is the unanimous accession of 40 clubs from Leagues 1 and 2.

This initiative proves to be useful because it provides points of comparison to clubs, while also enabling the creation of a positive image of the sector in relation to third parties, such as banks, suppliers, and sponsors. It also provides an economic perspective, by obtaining either accurate financing or sponsorship information. Such experience could serve as a catalyst to remove financial control from the "clubby" power, which is still not transparent about its business in football, thus facilitating all sorts of illegal practices. Another solution would be to structure regulation of sport, such that regulators of the sector shall have enough power to verify the accountability of clubs.

In Brazil, the CNE does not have the same authority as the DNCG, because its actions are limited to the establishment of guidelines, without powers to conduct more individualized analysis of the economic situation of clubs, much less stop financial negotiations. The creation of a body similar to DNCG in Brazil could constitute an important message that signals the repudiation of malpractices. It should comprise a totally independent structure. It is important to note that, in its composition, not only do people from the sport sector appear but so do members of the community, including trade unions, associations, auditors, and others.

3.5   The Need for Suspicious Activities Reports

The principle of confidentiality cannot be invoked for neglecting suspicious activities reports in the wake of FATF Recommendation No. 9. Indeed, the duty of professionals not connected with the financial sector to report, also under the FATF Recommendations Nos. 18, 21, and 22, is an essential tool to combat misuse of good practices by managers in hiring players. In short, the legislation that advocates for autonomy regarding the organization and functioning of sports bodies cannot neglect to require effective financial and administrative transparency, as well as civil and criminal liabilities of its directors.

Football is not only a sport, it is treated as a cultural heritage around the world. Football is consecrated in Brazil as a lively and inventive dance under the applause of fans. It includes strong body language, with musical and sentimental connotation that has cultural recognition worldwide. For Leonardo Schmitt Good and Rafael Teixeira Ramos, "sport in today's society is imbued with a transcendental aspect without paradigm and any other matters of human existence."[47] All of this makes it worthy of specific protection. Being a part of the cultural heritage, with specific

---

[47] *In*: Autoria coletiva. *Direito Desportivo. Tributo a Marcício Krieger.* Coordenação de Leonardo Schmitt de Bem e Rafael Teixeira Ramos. São Paulo: Quartier Latin, 2009, p. 18.

constitutional protection, football requires a particular treatment, the same given to the universal culture.

The social concern with cultural manifestation relies on what is best in human beings: our explosion of expression. It is essential to recognize the benefits that culture represents: evoking feelings in people, mainly of reflection, patriotism, and pleasure, and sometimes reconciliation and generosity. This makes the emotions coming out of a football game analogous to the emotions felt after viewing a work of art. However, similar to other works of art, football is often an environment for money laundering.

The United Nations Declaration of Human Rights of 1948 enshrines this subject, as seen in the following:

> Article XXVI.
> 1. Everyone has the right to education. Education shall be free, at least in the elementary and fundamental stages. Elementary education shall be compulsory. Technical and professional education shall be made generally available and higher education should be equally accessible to all on the basis of merit.
> 2. Education shall be directed to the full development of the human personality and to the strengthening of respect for human rights and fundamental freedoms. It shall promote understanding, tolerance and friendship among all nations, racial or religious groups, and shall further the activities of the United Nations for the maintenance of peace.(…)
> Article XXVII.
> 1. Everyone has the right to freely participate in the cultural life of the community, to enjoy the arts and to share in scientific advancement and its benefits.
> (…)

On the other hand, Brazil, as a signatory to the United Nations Educational, Scientific and Cultural Organization (UNESCO) Convention on the Protection of the World Cultural and Natural Heritage of November 16, 1972 (promulgated by Decree 80978 on December 12, 1977), has a duty to cherish and pass on to future generations cultural heritage, including "works of man" (Articles 1 and 4). Under that Convention, cultural and natural heritage should be given a function in the life of the community (Article 5).

Therefore, it is the duty of all to preserve the cultural heritage of humanity, as envisaged in the mentioned Convention, the result of the General Conference of UNESCO Educational, Scientific and Cultural Organization, meeting in Paris on November 21–23, 1972 (approved by Legislative Decree 74 of June 30, 1977).

It seems here that the values that are used to defend and protect unique work and behavior, by its magnitude and beauty, allow several people to enjoy the beauty and have moments of happiness. It would be complete nonsense, especially because of the cultural and pedagogical function of vital importance, to fail to recognize the manifest public interest in having several regulatory tools in the sport sector. The art of football should be protected from private appropriation under this manifest public interest.

By having a cultural nature that manifests social values, football deserves proper treatment in the wake of the very same legislation that enshrines the protection of cultural, historical, and artistic assets. This thought process allows for legislation on money laundering that includes suspicious activities reports of individuals, companies, and entities entrusted with the promotion, brokerage, and trading rights transfer athletes.

Bibliography

ANCEL, Marc. A nova defesa social: um movimento de política criminal humanista. Transl. Osvaldo Melo. Rio de Janeiro: Forense, 1979.

ANDRADE, Manuel da Costa. A nova lei dos crimes contra a economia (Dec.-lei n. 28/84, de 20 de janeiro) à luz do conceito de bem jurídico. In: CORREIA, Eduardo et al. Direito penal econômico e europeu: textos doutrinários. Vol. 1. Coimbra: Coimbra Ed. 1998. vol. 1.

_____; Costa, José de Faria. Sobre a concepção e os princípios do direito penal econômico. In: CORREIA, Eduardo et al. Direito penal econômico e europeu: textos doutrinários. Coimbra: Coimbra Ed., 1998. vol. 1.

_____. A Vítima e o Problema Criminal. Coimbra: Coimbra, 1980.

_____; Dias, Jorge de Figueiredo. Problemática geral das infrações contra a economia nacional. In: CORREIA, Eduardo et al. Direito penal econômico e europeu: textos doutrinários. Coimbra: Coimbra Ed., 1998. vol. 1.

ANDREUCCI, Ricardo. O direito penal máximo. Revista da Associação Paulista do Ministério Público. no. 35. pp. 48–49. São Paulo: Associação Paulista do Ministério Público, Oct–Nov 2000.

ARAÚJO Jr., João Marcello de. O direito penal econômico. Revista Brasileira de Ciências Criminais. vol. 25. pp. 142–56. São Paulo: Ed. RT, Jan–Mar 1999.

ARRUDA, Eduardo. A ameaça. Folha de São Paulo, Esporte, D2, Painel FC, March 31, 2010.

ASCENSÃO, José de Oliveira. Branqueamento de Capitais: reacção criminal. Estudos de direito bancário. Coimbra: Coimbra, 1999.

AUDIÊNCIA na França faz Fifa ameaçar país. Ex-técnico e ex-presidente da federação francesa vão à Assembleia Nacional hoje dar explicações sobre fiasco. _Folha de São Paulo_, Copa 2010, D18, June 6, 2010.

AUSTRÁLIA é acusada de subornar a Fifa para receber a Copa-2022. Federação deu colares de pérolas a comitê que escolherá a sede. Folha de São Paulo. Copa 2010, D13, July 1, 2010.

BEDÊ Jr, Américo et al. _Garantismo penal integral_. Salvador: JusPodivm/Escola Superior do Ministério Público da União, 2010.

BEM, Manuel Schmitt de; TEIXEIRA RAMOS, Rafael. Autoria coletiva. Direito Desportivo. Tributo a Marcício Krieger. Coordenação de Leonardo Schmitt de Bem e Rafael Teixeira Ramos. São Paulo: Quartier Latin, 2009.

BITENCOURT, Cezar Roberto. Manual de direito penal: parte geral. 6. ed. São Paulo: Saraiva, 2000. v. 1.

BRANCO, Vitorino Prata Castelo. _A defesa dos empresários nos crimes econômicos_. São Paulo: Saraiva, 1982.

BROYER, Philippe. La nouvelle économie criminelle. Criminalité financiére—comment Le blanchiment de l'argent sale et le financement du terrorisme sont devenus une menace pour lês entreprises et les marchés financiers. Paris: Éditions d'Organisation, 2002.

BUSTOS RAMÍREZ, Juan. _Manual de derecho penal español_: parte general. Barcelona: Ariel, 1984.

_____; LARRAURI, Elena. La imputación objetiva. Santa Fé de Bogotá/Colômbia: Temis, 1998.

CALLEGARI, André Luís. _Direito penal econômico e lavagem de dinheiro_: aspectos criminológicos. Porto Alegre: Livraria do Advogado, 2003.

_____. Importância e efeito da delinqüência econômica. _Boletim do Instituto Brasileiro de Ciências Criminais,_ São Paulo, np. 101, Apr 2001.

CAMARGO, Antonio Luis Chaves. _Imputação objetiva e direito penal brasileiro_. São Paulo: Cultural Paulista, 2002.

CAPEZ, Fernando. _Curso de Direito Penal—parte especial_. São Paulo: Saraiva, 2008.

CARVALHO, André Diniz de. A profissão de empresário desportivo—uma lei simplista para uma actividade complexa? Desporto e Direito, _Revista Jurídica do Desporto,_ Ano I, n. 2. Coimbra: Coimbra Editora, Jan./Apr. 2004.

CARVALHO, Márcia Dometila Lima de. *Fundamentação constitucional do direito penal.* Porto Alegre: Sérgio A. Fabris, Editor, 1992.

CASTILHO, Ela Wiecko V. de. O controle penal nos crimes contra o sistema financeiro nacional. Belo Horizonte: Del Rey, 1998.

CAVERO, Percy García. *Derecho Penal Económico—Parte General.* 2nd ed. Lima: Grijley, 2007.

CYR CLARKE, Natalie L. St. The Beauty and The Beast: Taming the Ugly Side of The People's Game. 17 Colum. J. Eur. L 601, 2010–2011.

COLON, Leandro; RIZZO, Marcel. Havelange, 96, renuncia a cargo na Fifa para não sofrer punição. Folha de São Paulo, Esporte, D4, May 1, 2013.

COMPARATO, Fábio Konder. Crime contra a ordem econômica. *Revista dos Tribunais.* São Paulo, vol 734, Dec 1996.

CORREA, Luís Fernando. A história do doping nos esportes. In: http://g1.globo.com/Noticias/ Ciencia/0,,MUL1267929–5603,00-A+HISTORIA+DO+DOPING+NOS+ESPORTES.html, August 16, 2009, accessed June 9, 2013.

CORREIA, Eduardo. Introdução ao direito penal econômico. In: CORREIA, Eduardo et al. *Direito penal econômico e europeu:* textos doutrinários. Vol. 1. Coimbra: Coimbra Ed., 1998. pp. 293–318.

_____. Introdução ao direito penal econômico. *Revista de Direito e Economia,* no. 3, 1977.

_____. Novas críticas à penalização de atividades econômicas. In: CORREIA, Eduardo et al. *Direito penal econômico e europeu:* textos doutrinários. Vol. 1. Coimbra: Coimbra Ed., 1998. pp. 365–373.

DELMAS-MARTY, Mireille, Sobre a concepção e os princípios do Direito Penal Económico. Direito Penal Económico e Europeu: Textos Doutrinários, vol. I.

_____. *Droit penal des affaires.* 3. ed. Paris: Presses Universitaire de France, 1990, t. 1.

_____; GIUDICELLI-DELAGE, Geneviève. *Droit penal dês affaires.* 4 ed. Paris: Presses Universitaire de France, 2000.

DIAS, Jorge de Figueiredo. Breves considerações sobre o fundamento, o sentido e a aplicação das penas em direito penal econômico. In: CORREIA, Eduardo et al. *Direito penal econômico e europeu:* textos doutrinários. Vol. 1. Coimbra: Coimbra Ed., 1998. pp. 374–386.

_____. Questões fundamentais do direito penal revisitadas. São Paulo: Ed. Revista dos Tribunais, 1999.

_____. Direito Penal. Parte Geral. Questões fundamentais da doutrina geral do crime. Coimbra: Coimbra Ed., 2004.

DIAS, José Carlos. Evasão de Divisas. In: ANTUNES, Eduardo Muylaert (Coord.). *Direito penal dos negócios*: crimes do colarinho branco. São Paulo: Associação dos Advogados de São Paulo, 1990.

DIAS, Maria Berenice. A Lei Maria da Penha na justiça: a efetividade da Lei 11.340/2006 de combate à violência doméstica e familiar contra a mulher. São Paulo: Ed. Revista dos Tribunais, 2007.

DIPP, Gilson. Legislação atrapalha o combate à lavagem de dinheiro [Interview on 11/03/2004]. *Consultor Jurídico.* www.conjur.com.br. Accessed June 18, 2012.

_____. Lava-jato de dinheiro. Entrevista. *Revista Época,* São Paulo, Oct. 28, 2004.

DOMENECH culpa imprensa por crise. Segundo deputado, treinador francês perdeu controle da equipe após jornal pôr na capa palavrões de Anelka. *Folha de São Paulo,* Copa 2010, D21, July 1, 2010.

EM Rondônia, concentração dura até a bola parar. Folha de São Paulo, Esporte, D11, 14 Mar. 2010.

EZABELLA, Felipe Legrazie, *Agente FIFA e o Direito Civil Brasileiro.* São Paulo: Quartier Latin, 2010.

FATF, Money laundering through the football sector, available at http://www.fatf-gafi.org/media/ fatf/documents/reports/ML%20through%20the%20Football%20Sector.pdf, accessed Feb 4, 2014.

FELTRIN, Sebastião Oscar. As ansiedades do juiz. *Revista dos Tribunais,* ano 77, vol. 628, pp. 275–78, Feb. 1988.

FERRAJOLI, Luigi. *Derechos y garantías—La ley del más de débil*. Transl. Perfecto Andrés Iba-
ñez. Madrid: Trotta, 1999.
_____. Derecho y razón. Teoría del garantismo penal. Madrid: Trotta, 1995.
_____. *Direito e razão. Teoria do garantismo penal*. Transl. Luiz Flávio Gomes et al. São Paulo:
Ed. RT, 2002.
_____. *El garantismo y la filosofía del derecho*. Bogotá: Universidade Externado de Colombia,
2000. Série de Teoria Juridica y Filosofia del Derecho, no. 15.
FIFA Code of Ethics, http://www.fifa.com/aboutfifa/organisation/footballgovernance/codeethics.
html, accessed June 1, 2013.
FIFA denied Spanish Courts to get a copy of the contract celebrated between the Brazilian football player Neymar and Barcelona club (in http://esporte.uol.com.br/futebol/ultimas-noti-
cias/2014/02/05/fifa-se-nega-a-enviar-contrato-de-neymar-para-justica-da-espanha.htm,   accessed Feb. 06, 2014).
FIFA exigiu mudanças na justiça da África do Sul. *Jornal da Globo*, TV Globo, June 23, 2010,
http://g1.globo.com/jornal-da-globo/, accessed June 24, 2010.
FIFA, Transfer Matching System. In http://www.fifa.com/aboutfifa/organisation/footballgover-
nance/transfermatchingsystem.html, accessed June 20, 2013.
FIFA Regulations Players' Agents. In http://www.fifa.com/mm/document/affederation/administra-
tion/51/55/18/players_agents_regulations_2008.pdf, accessed July 1, 2013.
FIFA Regulations on the Status and Transfer of Players, http://www.fifa.com/mm/document/af-
federation/administration/01/95/83/85//regulationsstatusandtransfer_e.pdf, accessed June 21,
2013.
_____. Estatuto da FIFA: http://www.fifa.com, accessed on May 05, 2013.
FINANCIAL Action Task Force—FATF. Money laundering through the football sector. In: http://
www.fatf-gafi.org/media/fatf/documents/reports/ML%20through%20the%20Football%20
Sector.pdf, accessed May 2, 2013.
FISCHER, Douglas. *Inovações no Direito Penal Econômico: contribuições criminológicas, políti-
co-criminais e dogmáticas*. Organizador: Artur de Brito Gueiros Souza. Brasília: Escola Supe-
rior do Ministério Público da União, 2011.
FRAGOSO, Heleno Cláudio. *Lições de direito penal: a nova parte geral*, 8ª ed. Rio de Janeiro:
Forense, 1985.
GARCIA, José Angel Brandariz. *El delito de defraudación a la seguridad social*. Valencia: Tirand
lo Blanch, 2000.
GENOVÊS, Vicente Garrido: MIR, Ricardo Sanches. Delincuencia de cuella blanco. Madrid: In-
stituto de Estudos de Política, 1987.
GODOY, Arnaldo Sampaio de. Direito tributário comparado e tratados internacionais fiscais. FA-
BRIS Sergio Antonio. Porto Alegre. p. 85, 2005.
GONÇALVES, Wagner. Ética na justiça: atuação judicial da advocacia pública e privada. *Etical*:
ética na América Latina. http://www.etical.org.br. Accessed May 9, 2005.
_____. Lavagem de dinheiro: conflito de competência da Justiça Federal. *Boletim dos Procura-
dores da República*, ano 4, no. 42, pp. 29–31, Oct. 2001.
GREENFIELD, Steve; OSBORN, Guy. The Football (Offences and Disorder) Act 1999: Amend-
ing s3 of the Football Offences Act 1991. J. C. L. 55 2000, p. 62, accessed April 1, 2013.
HASSEMER, Winfried. *Fundamentos del derecho penal*. Transl. Muñoz Conde. Barcelona:
Bosch, 1981.
_____; MUÑOZ CONDE, Francisco. *La responsabilidad por el producto en derecho penal*. Va-
lencia: Titant lo Blanch, 1995.
HEEM, Virginia, and HOTTE, David. *La lutte contre le blanchiment des capitaux*. Paris: Librarie
Générale de Droit et de Jurisprudence (LGDJ), 2004.
HUNGRIA, Nelson. Comentários ao Código Penal: arts. 1 a 10, 11 a 27, 75 a 101. 4. ed. Rio de
Janeiro: Forense, 1958.
IRVING, James G. Red Card: The Battle Over European Football's Transfer System. 56 U. Miami
L. Ver. 667, 2001.

JAKOBS, Günther. *Atuar e omitir em direito penal*. São Paulo: Damásio de Jesus, 2004. Série Perspectivas Jurídicas.

_____; MELIÁ, Manuel Cancio. *Direito penal do inimigo. Noções e críticas*. Transl. André Luís Callegari and Nereu José Giacomolli. Porto Alegre: Livraria do Advogado, 2005.

JENNINGS, Andrew. A ginga perfeita dos donos da bola. A FIFA controla o dinheiro, marca os adversários e dribla a Justiça. Entrevista. *O Estado de São Paulo*, J4, aliás, em 27.06.2010.

JESCHECK, Hans-Heinrich. *Tratado de derecho penal—Parte general*. 4th ed. Transl. José Luis Manzanares Samaniego. Granada: Comares, 1993.

JOHANSSON, Jens; MADSEN, Lars Backe. On the 'muscle drain' and (child) trafficking and football. Den Forsvunne Diamanten, Tiden Norsk Forlag, Noruega, Oct. 2008.

JOHNSON, Graham. Football and Gangsters. How organized crime controls the beautiful game. Great Britain: Cox and Wyman Ltd., 2007.

LARRAURI, Elena. *La imputación objetiva*. Santa Fé de Bogotá/Colômbia: Temis, 1998.

LIAO, Ricardo, Artigo publicado em "Legislação". Gazeta Mercantil, Nov. 28, 2002.

LIGAÇÃO com o Corinthians faz FPF tirar juiz da final. Paulista. Empresa de Rodrigo Braguetto tem contrato com o clube há dois anos. *Folha de São Paulo*, D2 esporte, May 18, 2013.

MACKENZIE, Simon, *Going, Gone: Regulating the Market in Illicit Antiquities*. In: http://www.mcdonald.cam.ac.uk/projects/iarc/culturewithoutcontext/issue18/gerstenblith_mackenzie_review.htm, accessed May 1, 2013.

MARTINS, Charles Emil Machado. A reforma e o "poder instrutório do juiz". Será que somos medievais? www.mp.rs.gov.br/areas/criminal/arquivos/charlesemi.pdf. Accessed Nov. 3, 2011.

MELO FILHO, Álvaro.*O Novo Direito Esportivo*. Brasília, São Paulo: Cultural Paulista, 2002.

MIGUEL, Ricardo Georges Affonso. A possibilidade de contratação do atleta menor de futebol e a utilização do instituto da antecipação de tutela para transferência do atleta de futebol. *Revista do Tribunal Regional do Trabalho da 1ª Região*, Rio de Janeiro, v.21, n.47, pp. 103–16, jan./jun. 2010.

MUÑOZ CONDE, Francisco. Principios politicocriminales que inspiran el tratamiento de los delitos contra el orden socioeconómico en el proyecto de Código Penal Español de 1994. Revista Brasileira de Ciências Criminais, São Paulo, no. 11, pp. 7–20, July/Sept 1995.

MUSCATIELLO, Vicenzo Bruno. Associazione per delinquere e riciclaggio: funzione e ilimite della clausola di reserva. Rivista Trimestrale di Diritto Penale Dell'Economia. no. 1. pp. 97–156. Padova: Cedam, Jan–Mar 1996.

NELSON, Timothy G. Flag on the Play: The Ineffectiveness of Athlete-Agent Laws and Regulations—and How North Carolina Can Take Advantage of a Scandal To Be A Model For Reform. 90 N.C. L. Rev. 800, 2011–2012, p. 803, April 1, 2013.

NORTE-AMERICANOS se rendem ao soccer e querem outra Copa. *Folha de São Paulo*, Copa 2010, D21, June 26, 2010.

OLIVEIRA, Eugênio Pacelli de, Coordinator. Direito e processo penal na justiça federal: doutrina e jurisprudência. São Paulo: Atlas, 2011.

OS EUA aprenderam a gostar de futebol. *O Estado de São Paulo*, E7, June 27, 2010.

PAGLIUCA, José Carlos Gobbis. A imputação objetiva (quase) sem seus mistérios. *Revista da Associação Paulista do Ministério Público*. São Paulo: ano IV, n. 35, out./nov./00. p. 35.

PEDRAZZI, Cesare. O Direito Penal das Sociedades e o Direito Penal Comum. *Revista Brasileira de Criminologia e Direito Penal*. Rio de Janeiro: Instituto de Criminologia do Estado da Guanabara, 1965, vol. 9.

PIERANGELLI, José Enrique; in: *O Consentimento do Ofendido na Teoria do Delito*. São Paulo: Revista dos Tribunais, 1989.

PINTO, Frederico de Lacerda da Costa. Crimes econômicos e mercados financeiros. *Revista Brasileira de Ciências Criminais,* São Paulo, no. 39, pp. 28–62, July/Sept 2002.

PRADO, Luiz Regis. Direito Penal Econômico: ordem econômica, relações de consumo, sistema financeiro, ordem tributária, sistema previdenciário, lavagem de capitais, crime organizado. 4th ed. rev. São Paulo: Ed. Revista dos Tribunais, 2011.

_____. Direito Penal Econômico: ordem econômica, relações de consumo, sistema financeiro, ordem tributária, sistema previdenciário, lavagem de capitais, crime organizado. 4th ed. São Paulo: Ed. Revista dos Tribunais, 2011.

RACISMO poderá gerar até rebaixamento. FIFA. Federação cria pacote de propostas para combater discriminação, com punições financeiras e esportivas. *Folha de São Paulo*, esporte, D3, May 7, 2013.

REALE Jʀ, Miguel. Despenalização no direito penal econômico: uma terceira via entre o crime e a infração administrativa? *Revista Brasileira de Ciências Criminais*. Vol. 28. pp. 116–129. São Paulo: Ed. RT, Oct-Sept 1999.

_____. *Instituições de direito penal—Parte geral*. Vol. 1. Rio de Janeiro: Forense, 2004.

RODRIGUES, Anabela Miranda. Contributo para a fundamentação de um discurso punitivo em matéria fiscal. Direito Penal Económico e Europeu: textos doutrinários. Coimbra: Coimbra ed., 1999.

ROXIN, Claus. Derecho penal—Parte general—Fundamentos. La estructura de la teoría del delito. Vol. I. Madrid: Civitas, 2006.

_____. *Funcionalismo e imputação objetiva no direito penal*. Translation and Introduction by Luís Greco. Rio de Janeiro: Renovar, 2002.

_____. *Problemas fundamentais de direito penal*. 3rd ed. Transl. Ana Paula dos Santos Luís Natscheradetz (Textos I, II, III, IV, V, VI, VII & VIII), Maria Fernanda Palma (Texto IX) & Ana Isabel de Figueiredo (Texto X). Lisbon: Vega Universidade/Direito e Ciência Jurídica, 1998.

_____. Reflexões sobre a construção sistemática do direito penal. *Revista Brasileira de Ciências Criminais*. Vol. 82. pp. 24–47. São Paulo: Ed. RT, 2010.

_____. *La teoria del delito en la discusión actual*. Transl. Manuel Abanto Vásquez. Lima: Grijley, 2007.

RUSSI, Joyce. ENCLA 2006. Entidades buscam aperfeiçoamento normativo. Formular leis que garantam o efetivo combate à lavagem de dinheiro e a recuperação dos ativos é a principal meta da Encla para este ano. *Jornal da Associação Nacional dos Procuradores da República*, no. 34, Feb. 2006.

SÁNCHEZ, Jesús-María Silva. *Eficiência e direito penal. Coleção Estudos de Direito Penal*. No. 11. São Paulo: Manole, 2004.

SANTOS, Cláudia Cruz. O crime de colarinho branco (da origem do conceito e sua relevância criminológica à questão da desigualdade na administração da justiça penal). Coimbra: Coimbra Ed., 2001.

SANTOS, Gérson Pereira dos. *Direito penal econômico*. São Paulo: Saraiva, 1981.

SECOND Report on the Situation of Human Rights Defenders in the Americas. Inter-American Commission on Human Rights. Published by Organization of American States—OAS, December 31, 2011.

SECURITIES and Exchange Commission—SEC: http://www.sec.gov/about/whatwedo.shtml, accessed May 2, 2013.

SIEKMANN, Robert. Labour Law, the Provision of Services, Transfer Rights and Social Dialogue in Professional Football in Europe. 4 ESLJ 1 2006–2007.

SILVA SÁNCHEZ, Jesús-María. *A expansão do direito penal. Aspectos da política criminal nas sociedades pós-industriais*. Transl. Luiz Otavio de Oliveira Rocha. São Paulo: Ed. RT, 2002.

TAX Authorities Provided Information to Football Money Laundering Report. HM Revenue and Customs has confirmed it gave details of two tax evasion cases. *The Guardian*, July 2, 2009.

TIEDEMANN, Klaus. Poder económico y delito (Introducción al derecho penal económico y de la empresa). Barcelona: Ariel, p. 33–34. 1985.

UNODC REPORT. Promoting health, security and justice. In: www.unodc.org, accessed April 13, 2010.

Chapter 4
Gambling and Lotteries

4.1 Initial Considerations

The sport and gambling industries are attractive sectors for the practice of money laundering due to the large monetary transactions involved and the growing number of people participating in them. Isolated, uncoordinated, and purely economic solutions are not enough to tackle the problem. We must work to uncover any legislative gaps that provide mobility, strength, and continuity to organized crime and enable unprecedented illicit wealth. The complexity of the sport and gambling sectors, along with the emotional involvement of the participants in these sectors, make it easier to succumb to the authorities in these fields, who deserve particular attention. Left unchecked, the problems caused by these authorities can lead to conflicts and instability with serious risk to the involved industries.

It is not possible to enable persistent tolerance of criminal practices in the sport and gambling sectors "in the name of sport." Doing so will eventually annihilate these sectors. Instead, enforcing property law and best practices in gambling and sport will preserve their credibility.

This study aims to understand the fragility of the situation of casinos and lotteries, a fact that is not as well known by the public. While the public wants the authorities to take action against money laundering and organized crime, it needs an updated perception of this world. Relying on outdated notions has allowed agents of organized crime to continue obtaining illicit riches through the perpetuation of a number of serious crimes. It is important to get answers that facilitate effective prosecution of these agents. For this reason, a sharp criminal intervention by the state is required at the outset, one that includes the forfeiture of unlawfully obtained goods and valuables.

Created in December 1989 by the seven richest countries in the world (G-7[1]), the Financial Action Task Force (FATF, or *Groupe d'Action Financière sur le*

---

[1] USA, Japan, Germany, France, UK, Italy, and Canada, which has since been joined by Russia (G8).

F. M. De Sanctis, *Football, Gambling, and Money Laundering,*
DOI 10.1007/978-3-319-05609-8_4, © Springer International Publishing Switzerland 2014

blanchiment des capitaux—GAFI[2]), organized under the aegis of the Organization for Economic Co-operation and Development (OECD), has a mandate to examine, develop, and promote policies for the war on money laundering. It initially included 12 European countries, along with the USA, Canada, Australia, and Japan. Other countries joined afterward (including China in 2007), as well as international organizations (including the European Commission and the Gulf Cooperation Council). Brazil joined, initially as an observer and later as a full member, at the 11th Plenary Meeting, held in September 1999.

The following Recommendations from FATF are relevant provisions contained in the 2012 version:

- Countries should identify, assess, and understand the money laundering and terrorism financing risks for the country and take action to mitigate them (*risk-based approach—RBA*, Recommendation No. 1).
- Countries should ensure cooperation among policy makers, the financial intelligence units (FIUs), and law enforcement authorities, and coordinate prevention and enforcement policies domestically (Recommendation No. 2). The current text of Recommendation No. 2 (previously contained in Recommendation No. 31) adds legitimacy to Brazil's National Strategy for the Fight against Corruption and Money Laundering (ENCCLA).[3]

---

[2] The FATF is an intergovernmental agency organized to promote measures for the fight against money laundering. Its list of Forty Recommendations, drafted in 1990, was revised in 1996. Another eight Recommendations were drawn up in 2003 (on financing of terrorism) and a ninth in 2004 (also about financing of terrorism). On February 16, 2012, all 49 Recommendations were revised, improved, and condensed into 40. These Recommendations are not binding, but they do exert strong international influence on many countries (including nonmembers) to avoid losing credibility, because they are recognized by the International Monetary Fund and the World Bank as international standards for combating money laundering and the financing of terrorism. In the 1996 version, they were adopted by 130 countries. In the 2003–2004 version, they were adopted by over 180 countries. It is important to mention that the idea of improving and condensing the Recommendations to avoid distortion and duplication, and to also incorporate the nine Special Recommendations on the financing of terrorism into the basic text (Forty Recommendations), originated in Brazil when it presided over the FATF between 2008 and 2009.

[3] According to a study conducted by the Brazilian Federal Justice Council's Judiciary Studies Center on the effectiveness of Law No. 9613/1998, through September of 2001, the Brazilian Federal Police had conducted only 260 police investigations, and most (87%) of the federal judges polled in that study answered that there were no active proceedings in their courts relating to money laundering through 12/31/2000, the date on the survey form (FEDERAL JUSTICE COUNCIL, *A critical analysis of the money laundering law*). In 2002 and 2003, with Minister Gilson Dipp of the appellate court presiding, and participation from representatives of the federal courts, the office of the federal prosecutor, the federal police and the Brazilian Federation of Bank Associations (FEBRABAN), the council drew up substantive recommendations to improve investigation and prosecution of criminal money laundering by engaging the cooperation of various government departments responsible for implementing the law. It was embryonic to the National Strategy for the Fight against Money Laundering and Recovery of Assets (ENCLA), later renamed the ENCCLA. The ENCCLA is made up of the primary agencies involved in the matter, which are the office of the attorney general, the Council for Financial Activities Control (COAF), the Justice Ministry's Asset Recovery and International Legal Cooperation Council Department (DRCI), the Federal Justice Council (CJF), the office of the federal prosecutor (MPF), the office of the

- The crime of money laundering should apply to predicate offenses, which may include any of a long list of serious offenses or any offenses punishable by a maximum penalty of more than 1 year, and criminal liability should apply to all legal persons, irrespective of any civil or administrative liabilities (Recommendation No. 3).
- No criminal convictions should be necessary for asset forfeiture. Furthermore, with reference to the Vienna Convention (1988), the Terrorist Financing Convention (1999), and the Palermo Convention (transnational organized crime, 2000), the burden of proof on confiscated goods should be reversed (Recommendation No. 4).
- Countries should criminalize the financing of terrorism (Recommendation No. 5).
- Countries should implement financial sanction regimes to comply with UN Security Council resolutions regarding terrorism and its financing (Recommendation No. 6).
- Countries should implement financial sanction regimes to comply with UN Security Council resolutions regarding the proliferation of weapons of mass destruction and its financing (Recommendation No. 7).
- Countries should establish policies to supervise and monitor nonprofit organizations in order to obtain real-time information on their size, activities, and other important features such as transparency, integrity, and best practices (Recommendation No. 8).
- Financial institution secrecy laws, or professional privilege, should not inhibit the implementation of the FATF Recommendations (Recommendation No. 9).
- Financial institutions should undertake customer due diligence (CDD) and verify the identity of the beneficial owner, and they should be prohibited from keeping anonymous accounts or those bearing fictitious names (Recommendation No. 10).
- Financial institutions should maintain records for at least 5 years (Recommendation No. 11).
- Financial institutions should closely monitor politically exposed persons (PEPs),[4] i.e., persons who have greater facility to launder money, such as politicians in high posts and their relatives (Recommendation No. 12).

---

comptroller-general (CGU), and the Brazilian Intelligence Agency (ABin), annually setting policy for all actions to be carried out in the execution of Law No. 9613/1998, on account of private and uncoordinated—if not conflicting—agendas having been observed among government agencies responsible for said enforcement. A meeting was held on December 5–7, 2003, in Pirenópolis in the State of Goiás, to develop a joint strategy for the fight against money laundering. To monitor progress toward the goals set forth in the objectives of access to data, asset recovery, institutional coordination, qualification and training, and international efforts and cooperation, an Integrated Management Office for the Prevention of and Fight against Money Laundering (GGI-LD) was created in compliance with Target 01 of ENCLA/2004. This office is comprised of the primary government agencies, as well as the Judicial Branch and attorney general's office, conducting both workshops and plenary meetings on various occasions. Every year, they define new actions (formerly targets), in hopes that the conclusions arrived at during their work sessions will be transformed into substantive outcomes.

[4] The 2012 version expanded the definition of PEPs to include both nationals and foreigners, and even international organizations.

Other provisions worth mentioning include the following:

- Financial institutions should monitor wire transfers, ensuring that detailed information is obtained about the sender and the beneficiary, and prohibit transactions by certain people pursuant to UN Security Council resolutions, such as Resolution 1267 of 1999 and Resolution 1373 of 2001, for the prevention and suppression of terrorism and its financing (Recommendation No. 16).
- Designated nonfinancial businesses and professions (DNFBPs), such as casinos, real estate offices, dealers in precious metals or stones, attorneys, notaries, and accountants, should be able to report suspicious activity, while being protected from civil and criminal liability (Recommendation Nos. 18 through 22).
- Countries should take measures to ensure transparency and obtain reliable and timely information about the beneficial ownership and control of legal entities (Recommendation No. 24), including information regarding trusts, namely information about the settlors, trustees, and beneficiaries of trusts (Recommendation No. 25).
- FIUs should have timely access to financial and administrative information, either directly or indirectly, as well as information from law enforcement authorities, in order to fully perform their functions, which include analyzing suspicious statements about operations (Recommendations Nos. 26, 27, 29, and 31).
- Casinos should be subject to effective supervision and rules to prevent money laundering (Recommendation No. 28).
- Countries should establish the means for conducting freezing and seizure operations, even when the commission of the predicate crime may have occurred in another jurisdiction (such as another country), and they should implement specialized multidisciplinary groups or task forces (Recommendation No. 30).
- Authorities should adopt investigative techniques, such as undercover operations, electronic surveillance, access to computer systems, and controlled delivery (Recommendation No. 31).
- The physical transportation of currency should be restricted or banned (Recommendation No. 32).
- Proportionate and deterrent sanctions should be available for natural and legal persons (Recommendation No. 35).
- There should be international legal cooperation, pursuant to the Vienna Convention (international traffic, 1988), Palermo Convention (transnational organized crime, 2000), and Mérida (corruption, 2003; Recommendation No. 36).
- Countries should provide mutual assistance toward a quick, constructive, and effective solution (Recommendation No. 37), including the freezing and seizure of accounts, even with no prior conviction (Recommendation No. 38), extradition (Recommendation No. 39), and spontaneously taking action to combat predicate crimes, money laundering, and terrorism financing (Recommendation No. 40).

In the 2012 revision, the Recommendations set forth general guidelines, with details given in interpretative notes. The interpretative notes fit within the context of common law and civil law, providing a common ground for countries with either legal system. In addition, the glossary makes it easy to place the adopted standards in proper perspective and provides important clarifications. One important innovation of the revised Recommendations, albeit not the purpose of the February 2012 review, was its emphasis on the need for countries to adopt the risk-based approach (RBA). Under RBA, countries must establish standards to guide public policies that address money laundering, terrorism financing, and the proliferation of weapons of mass destruction before applying measures that prevent and combat these problems.

Some Recommendations could have a special role for combating illegal gambling and money laundering. The FATF gave particular attention to DNFBPs, such as casinos and real estate offices, which must report suspicious operations (Recommendation Nos. 18 through 21). In addition, the FATF established a specific Recommendation directed toward casinos that subjects them to effective supervision and rules to prevent money laundering (Recommendation No. 28).

Large investments in casinos can create a real, positive economic impact when they are channeled with great social membership, business development, and extensive transmission of cultural values. However, the growth of this industry has been faced with criminal practices, notably corruption, tax evasion, and money laundering. While certain controls to stop money laundering have been put in place through the FATF guidelines, this has led to the search for new mechanisms to launder assets in order to unlink them to the predicate crime. There are obvious risks that arise when people use legitimate sectors for illicit gain, often leading to the contamination of these sectors with illicit money. Moreover, the global financial market and the development of information technologies have gradually strengthened the underworld economy, extending the possibilities of the practice of economic crimes.

Forms of gambling, such as casinos and lotteries, are occasionally the subject of discussion with respect to illicit financial crime. The study of such gambling activities is a paramount issue due to their vulnerability to criminal exploitation. For instance, Brazil received special attention due to the prevalence of the game "Bingo" in the country, which was created in order to stimulate sport before its alleged link with known clubs or federations.

Even certain court decisions, whether or not in favor of gambling, have demanded specific analysis regarding the remarkable possibility for money laundering that accompanies gambling.

Misha Glenny, in an important reflection, reveals that:

> But given that the shadow economy has become such an important economic force in our world, it is surprising that we devote so little effort to a systematic understanding of how it works and how it connects with the licit economy. This shadow world is by no means distinct from its partner in the light, which is itself often far less transparent that one might suspect or desire.[5]

This quote illustrates the importance of enforcement authorities paying special attention to dubious payments and constant movements of large sums of money. For example, without such attention, gambling houses can transfer or deposit funds through money changers or extra banking activities, thus preventing them from

---

[5] *McMafia: Crime sem fronteiras*. Trad. Lucia Boldrini. São Paulo: Companhia das Letras, 2008, p. 17.

being adequately controlled. Unfortunately, many countries have little experience in controlling this business practice, which may pose a high risk of money laundering. There would be neither an effective exchange of information between relevant authorities responsible for overseeing this business practice nor a clear definition of who would be responsible for sharing information.

This could, certainly, lead to suspicious transaction reporting of only one or a few isolated acts to the local FIU (in the USA, the Financial Crimes Enforcement Network, FinCEN; in Brazil, the Council for Financial Activities Control, COAF), which would only have limited effectiveness. Thus, the set of illicit practices, diluted with various chains of casinos, would not lead to knowledge of the entire illegal transaction, because it could only be verified by knowing about all of the illicit activity.

To combat the practice of economic and financial crimes, we also need to measure the problem and study the methods used to launder dirty money. Given the controls that are increasingly established and the ease in laundering money, gambling houses are constantly subject to exploitation by criminals through illegal control of operations or the purchase of their own establishments, often leading to larceny, fraud, and money laundering. In order for gambling houses to continue, it is essential to have customer confidence in the institution. This is why authorities must allow the honest play of games through an adoption of specific rules and require management to ensure a high standard of safety and supervision.

A great deal of attention has been focused on money laundering due to the highly sophisticated nature of its criminal practices. These practices have been internationally organized and professionally executed for a considerable amount of time. Organized crime has had a relatively free hand in its efforts to make criminal assets legal. This is made possible by the relative ineffectiveness of current national and international laws, which have not kept pace with the changing situation.

Gilson Dipp points out that organized crime takes advantage of the "inertia of states, and their closely regulated executive, legislative, and judicial branches, which are bound by the principle of territoriality—the idea that the law holds only within its boundaries. This is a hopelessly dated notion. Each state must, without giving up its sovereignty, achieve broad international cooperation. To insist on a nineteenth-century conception of sovereignty is to allow organized crime to exercise its will to the detriment of formal sovereignty."[6]

Francisco de Assis Betti views financial crimes as crimes that are generally "marked by the absence of social scrutiny, due to several factors including an excessive attachment to material things such as profit and egotistical zeal among the owners of capital, who are scornful of the lower classes and confident in their own impunity. Most of these crimes are covered up by collusive public officials. When the crimes do come to light, evidence is poorly produced and the facts are difficult to ascertain, given the specialized assessment required, culminating almost always in impunity."[7] Francisco de Assis Betti adds that it is not always "easy for a criminal

---

[6] Interview published Nov. 3, 2004, on the *Consultor Jurídico* website. www.conjur.com.br. Accessed June 18, 2012.

[7] BETTI, Francisco de Assis. *op. cit.*, p. 20.

to use the proceeds of crime. Profligate spending and the eccentricities that always accompany the easy acquisition of money, and immediate purchases way above one's standard of living, are outward signs of wealth which give rise to suspicion, and are conducive to investigations by either police or internal revenue authorities. Experienced criminals therefore try to come up with arrangements for investing their criminal proceeds and work with others inclined to conceal these assets and obliterate the money trails in order to avoid enforcement efforts."[8]

To the extent that society has realized that serious crime can encompass more than just violent crime, more and more states have ratified international regulatory instruments without restrictions, demonstrating that they are no longer willing to tolerate open-ended criminality within their borders. It should be noted that money laundering is in essence a derivative crime, because the offense is contingent upon an antecedent crime. This link between money laundering and organized crime necessitates immediate and aggressive intervention by governments to ensure the very survival of their countries.

One could indeed define money laundering as a simple procedure, whereby one transforms goods acquired through unlawful acts into apparently legal goods. However, overriding considerations of legality and legal security do not permit us to make use of such a simple definition. Another difficulty with money laundering is that it is not simple to accomplish, nor does it follow any preset rule. The commission of the crime involves processes that are often complex and sophisticated. Classically speaking, the crime of money laundering involves three stages of conduct: (1) concealment or placement, in which goods acquired by unlawful means are made less visible; (2) monitoring, dissimulation, or layering, in which the money is severed from its origins, removing all clues as to how it was obtained; and (3) integration, in which the illegal money is reincorporated into the economy after acquiring a semblance of legality. Added to this is the recycling stage, which consists of wiping out all records of the previously completed steps.

Faced with the complexity of the various forms of conduct and processes comprising money laundering, one is struck by the almost complete impossibility of imposing legal restraints, other than through combined means (i.e., proscribing more than one form of conduct) or open-ended means (i.e., targeting a large number of activities described in the Vienna Convention and adopted by most countries). Additionally, money laundering is always a derivative crime that is necessarily connected to its antecedent crime. All these issues add innumerable peculiarities to the crime of money laundering, peculiarities which must be gradually sorted out by jurisprudence or case law.

In Brazil's case, money laundering was not typified in the main body of the criminal code, as was done, for instance, in the USA (see 18 U.S.C. § 1956). This poses an undeniable difficulty. If the crime in question was codified, it would be promptly adapted to the principles and rules of the criminal code. Because the money laundering system is integrated and hierarchical, there would be no margin for unjustifiable exceptions. This is the case in France, Italy, Switzerland, and Colombia.

---

[8] BETTI, Francisco de Assis. *op. cit.*, p. 39.

The United Nations Convention against Transnational Organized Crime convened in Palermo on November 15, 2000,[9] following the United Nations Convention against Illicit Traffic in Narcotic Drugs and Psychotropic Substances of December 20, 1988[10] (Article 5). Both global regulatory guidelines require the state parties to make the laundering of the proceeds of crime a crime itself (Article 6), and they provide for the confiscation of "proceeds of crime derived from offences covered by this Convention or property the value of which corresponds to that of such proceeds" (Article 12(1)(a)). Parallel to that is the United Nations Convention against Corruption held at Mérida in 2003 (Article 31, item 5—confiscation and seizure of money in an amount equivalent to the proceeds of crime).[11]

Items 2, 3, and 4 of Article 12 of the United Nations Convention against Transnational Organized Crime held at Palermo correspondingly assert that: "State Parties shall adopt such measures as may be necessary to enable the identification, tracing, freezing or seizure of any item referred to in paragraph 1 of this article for the purpose of eventual confiscation; if the proceeds of crime have been transformed or converted, in part or in full, into other property, such property shall be liable to the measures referred to in this article instead of the proceeds; if proceeds of crime have been intermingled with property acquired from legitimate sources, such property shall, without prejudice to any powers relating to freezing or seizure, be liable to confiscation up to the assessed value of the intermingled proceeds." Such provisions accurately depict the new world order with respect to combating organized crime, including narcotics trafficking and corruption.

It is sometimes alleged by defendants that the property seized has no links to the crime. The judge must then properly estimate the amount that flowed from the proceeds of the unlawful conduct imputed, being mindful of the need to enforce the requirements set forth in the foregoing Conventions, as well as Article 387, Section IV, of the Brazilian Code of Criminal Procedure. This Article requires that the sentence be fixed at the "minimum amount required for reparation of damages caused by the infraction, taking into account all losses suffered by the aggrieved party," in order to secure definitive forfeiture of that amount to the injured party or to the state as indemnification for damages caused by unlawful conduct.

Under Article K.3 of the Treaty of Maastricht (1992), EU member states agreed to adopt a common policy in their domestic efforts, and the 1998 joint action (98/773/JHA) sought to include money laundering as a type of organized crime. This was revoked in part by the Framework Decision[12] of the European Union Council, dated

---

[9] The Convention against Transnational Organized Crime was promulgated in Brazil by Decree No. 5015, dated March 12, 2004, and passed by Legislative Decree No. 231, dated September 29, 2003.

[10] The Convention against Illicit Traffic in Narcotic Drugs and Psychotropic Substances was ratified in Brazil by Decree No. 154, dated June 26, 1991.

[11] The Convention Against Corruption was ratified in Brazil by Decree No. 5687, dated January 31, 2006.

[12] Decisions and framework decisions were new instruments under Title VI of the European Union Treaty ("Provisions on Police and Judicial Cooperation in Criminal Matters") replaced joint action. Framework decisions are used to bring together the legislative and regulatory provisions of

June 26, 2001. Under this decision, member states agreed to not make reservations on Articles 2 and 6 of the European Convention of 1990 (including the rule which provides for money laundering resulting from general criminal conduct), since only *serious infractions* can be at issue. The member states also provided measures for the confiscation of proceeds from crimes that either have a maximum penalty of greater than 1 year or are considered to be serious crimes (Article 1).

The Framework Decision of February 24, 2005 (2005/212/JHA), regarding the forfeiture of products, instruments, and property related to crime, allows "extended powers of confiscation" aimed at not only the forfeiture of the assets of those found guilty but also the assets acquired by their spouses, companions, or those whose property transferred to some company under the influence or control of the guilty parties. These extended practices apply to organized criminal practices, such as counterfeiting, trafficking of persons or assisting illegal immigration, sexual exploitation of children and child pornography, trafficking of narcotics, terrorism, terrorist organizations, and money laundering, as long as these crimes are punishable by a sentence of at least 5 years of imprisonment (or, in the case of laundering, a maximum penalty of at least 4 years of imprisonment), and they generate financial income (Article 3, Sections 1–3).

Note that the Palermo Convention provides for international cooperation on matters of confiscation (Article 13(1)) and expressly provides that the proceeds of crime be allocated to finance a United Nations Organizations Fund to assist member states in obtaining the wherewithal needed to enforce the Convention (Articles 14(3)(a) and 30(2)(c)). Any illegal proceeds can be included within the scope of this Convention if it can be shown by convincing evidence that they may be related to the commission of antecedent crimes and to money laundering. Thus, if gambling was indeed being used for purposes of money laundering, those circumstances would justify judicial search and seizure, and possibly confiscation, of gambling proceeds.

Leaving illegally obtained money in the hands of criminals—especially members of organized criminal gangs—encourages the reentry of these monies into the underworld or back into the original illegal business practices, creating the potential for serious harm to society. To prevent the use of the sports and gambling sectors for organized crime, it is important to have an adequate understanding of the methods used to launder illicit funds, the vulnerabilities of these methods, and the capacity to exchange vital information from foreign authorities. However, even assuming a high degree of sophistication, supervision, and safety, gambling houses can still be seen as attractive settings for crime to criminals who do not fear the consequences of their personal illegal actions.

---

member states. They are proposed on a motion by the commission or by a member state, and must be unanimously adopted. They are binding on member states as to results to achieve, and leave it to national courts to decide on the manner and the means of achieving them. Decisions address all other goals besides the conference committee work on legislative and regulatory provisions of the member states. Decisions are binding and all measures necessary to carry out the decisions within the scope of the EU are adopted by the council through qualified majority vote.

4.2 Casinos

The operation of casinos is very similar to the operation of other financial institutions because of the extensive payments with cash, the exchange of chips for cash or checks, and the frequent participation by foreign tourists, who are the constant beneficiaries of certain amenities like room and board.

This undeniable movement of people and resources requires strict transparency to prevent the use of casinos for organized crime. Casinos on cruise ships raise important questions regarding jurisdiction to prosecute any illicit activity: Whether jurisdiction is based on where the ship is registered or where it travels or operates. Countries do not have specific regulations regarding this issue, which may result in a lack of jurisdictional action that limits the ability to prosecute criminal activity.

In addition, there is the possibility of online gambling, which gives rise to a study of how to regulate this subset of gambling effectively and how to enforce official controls to prevent money laundering in this area. Moreover, there is no control over foreign arms within this subset of gambling.

The importance of the gambling sector can be measured by numbers. It is estimated that business at casinos generated more than US$ 70 billion in 2006. This does not include the profits flowing from the proliferation of Internet gambling sites, which generated more than US$ 11 billion in 2006.[13] According to Johan Warren Kindt and Stephen W. Joy, "[P]olicy-makers worldwide generally failed to identify the large socio-economic costs associated with Internet gambling, as well as the ability of Internet gambling and other forms of cyberspace gambling to destabilize local, national, and even international economies by disrupting financial institutions."[14] Even in countries that properly regulate the gambling industry, major money laundering still exists.

A March 2009 report from the international FATF[15] revealed that there are many suspicious activities reports related to the gambling sector. It is very easy to convert illicit cash through electronic or cash transactions in the gambling context. For example, it is possible to exchange illicit monies through "buy-ins" and "cash outs." In the first case, there is a conversion of money into tokens, tickets, or credits in order to start the game. In the second case, the reverse occurs, and tokens, tickets, or credits are replaced with casino checks, claims on accounts, or fund transfers to other casinos.

Likewise, it is possible to convert the ticket called "TITO" (Ticket-in/ticket-out), which allows a gaming machine to accept bills or tickets with credits printed on it

---

[13] Will U.S. go after online gambling? You can bet on it, Financial Post, Sept. 20, 2006, http://www.financialpost.com/story.html?id=56c42e74-32e5-43a7-af0b-1809acde2a09&k=66611.

[14] *See* John Warren Kindt and Stephen W. Joy, *Internet Gambling and the Destabilization of National and International Economies: Time for a Comprehensive Ban on Gambling Over the World Wide Web.* 80 Denv. U. L. Rev. 111, (2002–2003), p. 111.

[15] Cf. Financial Action Task Force—FATF *Vulnerabilities of Casinos and Gaming Sector,* March 2009. Available at http://www.fatf-gafi.org/media/fatf/documents/reports/Vulnerabilities%20of%20Casinos%20and%20Gaming%20Sector.pdf, Accessed June 10, 2013.

(i.e., a ticket-in) or print tickets containing credits once the player wishes to settle the game (i.e., a ticket-out). In this case, the customer can exchange the ticket for cash at the establishment or reuse it on another "TITO" machine to restart the game.

Brazil, despite being one of the most populous countries in the world with more than 200 million people, has a small industry of casinos. There were about 130,000 machines, including slot machines, in about 1,500 "Bingos" that operated across the country, reaching its peak in 2006. However, in 2007, many casinos were forced to a standstill, and many machines were confiscated by the authorities, when the practice of corruption in the industry was revealed by the federal police's "Hurricane" operation. Their findings indicated alleged involvement of politicians and organized crime within the gambling industry in order to keep the casinos open.

A bill to legalize casinos could create a serious precedent for the practice of money laundering if it does not come with a great structure for the supervision of gambling. Without strict regulation and sufficient mechanisms for oversight, it is not possible to keep organized crime away from this sector. Even if the gambling sector leads to more jobs and investments that benefit the country, legalizing gambling is not justified unless there is also an effective mechanism for preventing organized crime.

4.3   Lotteries

The lottery is a prominent gambling activity that is usually controlled by the government and is open to the general public. However, evidence of money laundering has already been detected through alleged lottery winnings by people who have obtained several awards through different modalities, earning millions in the process. There are even cases involving people winning several hundred awards. In one case, money collected through tax evasion was invested in the lottery and provided winnings totaling over 500 awards, where all the awards were divided between three people linked to a single commercial establishment. This indicates the faltering performance of communications within the FIUs (e.g., the FinCEN in the USA and the COAF in Brazil).

Unfortunately, when hearings are held regarding these awards, they often do not provide useful information, particularly when the winner states that the game was legitimately played. Often, prosecutors request clarifying information from the appropriate administration office (in Brazil, the National Superintendence of Lotteries and Games of the *Caixa Econômica Federal* or CEF), such as the drawing dates, amounts and types of lotteries uncovered by a particular investigation, the income tax paid, and the bank account agency which was paid for each award. The collection of this information is accomplished by presenting the original ticket that was used. For payments above a certain amount (in Brazil, R$ 800.00), two copies of a statement of accrued income (DAPLOTO) must be filled out—one is delivered to the customer and the other is archived for 5 years with the appropriate department of its administration.

Repeated awards in a short period of time defy the most rudimentary notions of mathematical probability. However, such results, although practically impossible, allow application of the principle *in dubio pro reo*. In the absence of nominal tickets, the modus operandi used for money laundering consists of the acquisition of award tickets from the intended real person, leading to the exchange of "dirty" money.[16] This is made possible by the involvement of game house owners—people who sell winning lottery tickets and are responsible for finding and enticing the winners, and even employees of the department responsible for its administration, to engage in this corrupt process.[17]

Another avenue for effective money laundering is created when people place bets in amounts that cover all the possibilities of success, which leads necessarily to less earned income, but allows illicit money used in games to be converted into legitimate earnings that are documented. Marcelo Batlouni Mendroni warns us about cases in which a person plays the requisite number of games to win by getting numerous slips of the game. The number of slips remains an unknown quantity because there is no identification or recording, and the person can earn a prize of high value, despite investing a greater amount than the prize itself. As a result, the money received would be clean to play.[18]

There are countries where clandestine games are classified as a criminal offense. For example, engaging in the "Animal Number Game" ("Jogo do Bicho") in Brazil is considered a criminal offense (Decreto-Lei, Decree no. 3688 of October 3, 1941), punishable by simple imprisonment of 4 months to 1 year and fine (Article 58). The punishment applies to both the operator of the game and the participant who plays to win the prize for himself or herself (or for third parties, in which case the penalty is limited to a fine).

Unlike the lottery, the businesses that operate the "Animal Number Game" do not pay taxes and there is, therefore, the allocation of social resources to those provided under legislation. The "Animal Number Game" is a popular lottery game; consequently, its bets have occurred freely in places that run on points without any identification, as well as in legitimate lottery houses, making the illicit bets difficult to detect. In many cases, any paper trail was replaced by the computer.

Historically, the "Animal Number Game" has been seen as a harmless offense. However, this is not the case. Motorcyclists take money from betting and disseminate the results of the draw, which takes place twice a day. Given this secrecy and collusion, corruption is likely taking place among public officials and politicians. Thus, this practice ultimately contributes to several other criminal acts that produce great social harm.

---

[16] Cf. Financial Action Task Force—FATF, Money Laundering through the Football Sector Report. Available at http://www.fatf-gafi.org/topics/methodsandtrends/documents/moneylaunderingthroughthefootballsector.html. Last updated Feb. 1, 2012. Accessed on May 10, 2013.

[17] According to statistical data compiled by the Brazilian FIU (COAF), the number of suspicious activity reports has been considerable—162,128 in 2011 and 195,493 in 2012. Since its inception in 1999, the COAF has received 637,666 reports, which shows that the possibility of money laundering through lotteries must be taken seriously. COAF, available at https://www.coaf.fazenda.gov.br/conteudo/estatisticas/comunicacoes-recebidas-por-segmento/-1, Accessed May 10, 2013.

[18] Cf. Marcelo Batlouni Mendroni, *Crime de Lavagem de Dinheiro*, p. 76.

4.4 Typologies (Money Laundering Methods)

The methods used to launder and use illicit assets are constantly evolving. For the standards to remain relevant and effective, researchers must keep up to date with the latest money laundering and terrorist financing methods, techniques, and trends. It is important to constantly monitor and identify new threats and risks to the financial system and to publish the findings in typologies studies. These studies are aimed at raising global awareness and facilitating early detection of the use and abuse of the systems. They are also instrumental in ensuring the development of the most appropriate standards to respond globally to these new and emerging money laundering and terrorist financing risks and threats. The conclusions generated by the typologies studies played an important role in the revision of the FATF Recommendations. The new FATF Recommendations, which were adopted in February 2012, provide countries with the tools to build stronger safeguards to face today's threats and challenges to the financial system.

There are records indicating the practice of jobbery, where moneylenders' exploitation is financed by organized crime. Through this practice, moneylenders convince customers in financial difficulty to not submit to legal loans and to instead obtain resources that support gambling. For example, the purchase of winning lottery tickets easily covers up "dirty" money by making it seem clean. Winning lottery tickets can also be obtained from bets in amounts that cover all the possibilities of success, allowing the conversion of illicit money to having a known and documented origin.

It is also worth mentioning the possibility of purchasing tokens or tickets through the use of credit cards, where the leftover tokens are exchanged for cash or casino vouchers. The casino tokens are considered valid instruments, most commonly issued for their use in slot machines. However, sometimes, credit cards are used to purchase narcotics, and the traffickers negotiate these deals in gambling houses. The company that manages the cards is paid with money received from the gambling houses. This mechanism allows for the illicit accumulation of wealth. Money launderers usually acquire chips or credits with cash or by depositing money in accounts with the gambling houses or casinos. In these cases, it is possible to use the credits or the gift certificates, known as chip purchase vouchers (CPVs), in casino networks in different countries. It greatly complicates the control over the casino system, as the possibly existing credit can be converted into a check-in setting among various casinos that are different from the first one that provided the chips or CPV.

Other hypotheses regarding illegal gains or the illegal use of gambling houses are also worth mentioning. Criminals launder counterfeit money by making use of agents who exchange money through multiple transactions made by anonymous people, using false documents created to disguise their illicit origin. For example, this exchange can occur by using chips as currency to conduct illegal transactions. Criminals can retain the chips for a period of time and use them to buy drugs or other illegal substances. These chips can be transported to other countries, serve as payments for clandestine activities, and eventually be exchanged in casinos by third parties in diluted amounts, which do not lead to any suspicious communication. These acts usually do not call attention to a particular gambling house unless it has a specific type of chip and does not allow the exchange of other types of chips, even though they come from the same network.

Criminals also launder illegal money by entering values into video poker machines, such as one, five, and ten dollar American bills, and then pressing the "cash out" button after playing briefly or not even playing at all, which generates a receipt that can serve as a document for a refund to present to the cashier. Another possibility lies in converting illicit money into legal money by buying chips for high prices, i.e., an inflated purchase price. The winner can aggregate cash and then exchange the total amount for casino checks. The purchase of award certificates that can be redeemed by or passed to others keeps some distance between the winnings and their illicit origins.

In Australia and Belgium, it should be noted that the purchase of accumulated money in chips occurs not to play with them, but to exchange their value through third persons linked to the buyer. In South Korea, the acquisition of chips using checks between 2003 and 2005 totaled US$ 20 million. Such chips were exchanged for cash and checks issued by casinos. The money was used for corrupt purposes by government officials.

In the USA, a lawyer in the state of New Jersey was convicted of accumulating over US$ 500,000 through fraud and laundering US$ 250,000 in a casino in Atlantic City. He transferred this amount to the casino and bought chips, playing for about an hour on roulette and losing US$ 10,000. He traded the rest for currency in cash and left the casino. A similar case occurred in Spain, where different people entered separately into a casino and obtained chips. After playing a few sums, they exchanged the chips for checks that were paid to a third person.

The report of the International Financial Task Force on laundering in football that took place in 2009[19] highlights the following indicators of money laundering: (1) inserting values in gaming machines and requesting their immediate exchange for credits; (2) seeking credits and not playing at all or playing very little; (3) trying to be friends with employees of gambling houses or casinos; (4) buying chips with little or no gambling; (5) using third parties to buy chips; (6) inconsistency between the amount of the bets and the customer's financial situation; (7) dramatic or rapid rise in size and frequency of transactions in a particular client account; (8) exchange of coins or paper currency notes for cash in the establishment; and (9) gaming machines, video lottery terminals (VLTs), and TITO machines (ticket-in/ ticket-out) are used to refine the currency through large sums, little gambling, and later-exchanged credits.

An interesting case involved the importation and distribution of heroin in Australia. The drug came from Vietnam. The person used large amounts of money and third parties to purchase chips in his name. On the same day, there was intensive

---

[19] Cf. Financial Action Task Force—FATF, Money Laundering through the Football Sector Report. Available at http://www.fatf-gafi.org/topics/methodsandtrends/documents/moneylaunderingthroughthefootballsector.html. Last updated Feb. 1, 2012. Accessed on June 10, 2013.

exchange of these chips for cash, careful to avoid exceeding AU$ 10,000 per transaction, the amount that triggers suspicious activity reports. Authorities discovered that there were some referrals by a remittance company to various entities in Vietnam, without the negotiator of the consignment connected to the "player."

There is also the possibility of minimizing suspicion by distributing large amounts of cash through small transactions to evade the legal limits that require communication with authorities in order to prevent laundering. This type of strategy is known as "structuring," which includes the following: (1) deposits or regular transactions below limits that require reporting to the authorities; (2) use of third parties to carry out transactions with single or multiple accounts; (3) use of regular checks from financial institutions to acquire tokens or chips, with each transaction being less than the limit that suggests a suspicious transaction and requires reporting; (4) requests to split awards in amounts below the legal limit and exchange them for cash at ATMs; (5) several people sending funds to a sole beneficiary; (6) checks issued to a player's relative; (7) inconsistent activities for the customer profile; (8) casino account transactions conducted by persons other than the account owner; (9) third parties who request structuring deposits and wire transfers; (10) large volumes of transactions in a small period of time; (11) high frequency of betting amounts that are always below the limit requiring reporting; (12) mismatch between the purchases and exchanges of chip currencies; (13) refusals of compliance with the use of third-party documents, whether they are false documents or those from tourists; and (14) suddenly straying from typical betting patterns for a particular account.

On the other hand, gambling house accounts, which are made available to customers for deposit and for converting credit lines, have fewer reporting requirements for suspicious transactions and thus allow an easy path to money laundering. For example, deposits made through electronic transfers can be used for cash or transferred to other accounts with little or no gaming activity. In such cases, the illegal money exchanger usually does not submit to obligations of reporting suspicious transactions and continues to have accounts that can supply money to casinos. Another suspicious signal of money laundering occurs when several people transfer funds to a single beneficial owner, followed by accountants or lawyers becoming in charge of these transactions. Some casinos offer safes for special customers. These safes present a serious risk because they lack transparency regarding the use by the customers or by third parties holding their passwords.

An interesting case occurred in which sums of money stemming from illicit drug trafficking, deceit, and credit card fraud traveled from England to Dubai after being laundered in a casino. Money played and exchanged provided the defendants an explanation as to the apparent lawful source of funds. In this case, the given explanation alleviated any raised suspicions almost instantly. These circumstances require more appropriate regulations to prevent suspicious transactions.

One money-laundering method used in lotteries in Brazil consists of offering award winners a premium on their earning in exchange for transferring the prize to criminals.

It also raises suspicion when two people in apparent opposition, but who are in fact engaged in collusion, place identical bets on the same game, when there is chance to win double or nothing (for example, in roulette: 1,000 red and 1,000 in black). This invariably allows one party to gain winnings, and that party issues a check to the other party without generating suspicion or notifying the authorities.

There is also the possibility of converting large amounts of currency into the currency of another country, which does not raise suspicion when there are a large number of foreign players, thus altering the original form of the currency. This method was used in one case in Spain, where a group of foreigners who separately bought chips in a casino using different currencies later converted their chips into euros. In this case, the casino not only detected the suspicious operation in advance, but it also ordered the operation's cancellation and reported it to the Spanish FIU.

There are also unusual cases that involve employees of gambling houses or casinos. For example, complicity among employees has led to a lack of reporting suspicious transactions, the destruction of documents related to such communication, and the falsification of players' data to justify the accumulation of credits. An important case illustrating this method came from the USA, where a group of drug trafficking money launderers bribed employees of a particular casino to access machines controlled by software that allowed the money launderers to take over certain features, thus enabling their illicit gains. Often, these types of illegal activities are made possible by contact between customers and employees outside the gambling houses.

There is also the possibility of money laundering through the use of stolen credit cards. However, it is easier for authorities to follow the trail of money with this method. In Belgium, for example, a person visited a casino on the country's coast on two occasions and acquired chips worth € 400,000 paid by cash and credit card. The casino reported the situation to the local FIU. The FIU verified that the account of that player was supplied by several transfers from companies and cash deposits and that the player's wife had business in Belgium, maintaining contacts with organized crime in Central and Eastern Europe. The defendant-player also maintained frequent contact with a person investigated for money laundering stemming from this organized crime.

Similarly, the use of debit cards is a valuable tool to commit fraud and money laundering. In England, it was possible to verify that a person acquired the maximum chips with a debit card without playing them, later exchanging them for cash. This caused the limitation of transactions with debit cards to a down limit for first operations. In addition to credit and debit cards, the use of false documents is another common method in gambling houses for opening accounts to conduct games and obtain winnings.

Several vulnerabilities have been identified by the FATF, but it is worth mentioning tourism activities related to casinos and gaming houses in particular, called "junkets." This is a marketing program that creates a tour organized specifically for gambling, which may include transportation, accommodation, incentives to play, and movement of funds to other casinos. It can be promoted by the casino itself or through outsourcing game houses. Participants in this type of tourism usually trust their operators to allow the movement of money across borders. This relationship between operators and customers has the potential of leading to complicity between them, thus enabling money laundering. The authorities should be notified of any suspicious transactions that occur in this setting as well. However, regional offices or outsourced casinos usually accept the previous deposits on tourists' behalf before the trip. These deposits sometimes occur by wire transfer, which do not call the attention of the local authorities that oversee money laundering.

To prevent misuse of junkets, their registration should be required before operation, with detailed qualification of authorized operators, including the requirement of filing fingerprints, so that they have the obligation to report suspicious transactions undertaken by customer-players. Moreover, the junkets should be subject to cancellation of their registration in cases of unlawful activity.

Gambling houses should also be required to report illegal junkets to the authorities. Often, junket activity is vital to the gambling houses, especially in sparsely populated countries, such that there is a very close relationship between the gambling house owners and junket operators. This may lead to the misuse of the junkets. Some junket operators may be able to gather a pool of customers, which can be used to mask individual spending. Plus, junkets in foreign countries that are not properly regulated enable connections to organized crime, even if they or companies related to them are not allowed to work in places where they have established gambling houses.

One cannot fail to mention that junkets may also be an alternative mechanism, formal or informal, for transfers of funds. The very nature of this activity has suggested that they are an informal mechanism. Some casinos also offer "junket agents" a nonnegotiable amount called "dead chips," which cannot be exchanged for cash or normal chips. These chips can be used as currency to facilitate criminal transactions. These aspects of junkets make it especially imperative to have regulatory requirements and an obligation to report suspicious activities.

There was an interesting case in which a casino's agent received large sums of money in China from a customer who wanted to play in Macao. This agent received the amount in a trade near Macao and divided the amount into parts that were physically transported to the island. All the money was deposited into an account of the casino's agent and then passed on to the gambling house. The gambling house converted the money to nonnegotiable chips. When the client had gained a certain amount of money, it was given to the agent who sent it back to China by unofficial routes.

A new issue has become the subject of concern: the growth of travel offers through casino ships with the system "junket," operated by independent operators. Normally, players deposit a significant amount of money with the junket operator. But the regulation of this kind of service is still lacking. One category of customers, the high-rolling players (or "high rollers"), occupies special VIP rooms in the complex and gets special treatment. This clientele is linked to junkets' business, and as such, they are vulnerable to potential identification and to the discovery of the origin and destination of their resources. Thus, the authorities have been hesitant to ease cash transactions, especially for high rollers. At the same time, the gambling houses have offered similar facilities to any financial institution, while the regulation and supervision of these entities is also not consistent.

Two examples illustrate the risk of money laundering that accompanies high rollers. First, in Australia, an Asian person linked to organized crime was considered a high roller and engaged in heroin trafficking through a casino-hotel, using gambling to mask illicit gains. He received incentives from the casino totaling AU$ 2.5 million and spent 2 years as a nonpaying guest. Second, in the USA, a foreigner used to take a trip to Las Vegas to gamble and lost approximately US$ 150 million. The casino offered a credit of US$ 10 million, plus other benefits such as hotel suites, cars, and aircraft services. The casino held wire transfers and bank accounts linked to the bettor on a corporation. There was no suspicious transaction report made by the casino to verify the source of funds.

To avoid such problems, it is essential to provide proper training for employees of gambling houses in order to prevent and detect money laundering. These employees must be certified and undergo training to report suspicious transactions. The untrained employee is prone to misconduct even when regulations are imposed. It is also essential to penalize violations of administrative rules (i.e., not implementing adequate internal controls to prevent money laundering) to create consensus and enforce these legal provisions.

In the USA, there are an estimated 567 Native American tribes recognized by the federal government (half of them in Alaska), and 223 of them operate gambling activities in about 28 states. In one case, the FinCEN decided to institute an administrative punishment to address the lack of implementation of preventive regulations by the Tonkawa Bingo and Casino and Edward E. Street, who operated it. The casino was located in Tonkawa tribe territory and self-governed by Native American tribes located in Oklahoma. The casino operated under the approval of the Tonkawa Tribal Gaming Commission. The casino's violations were based on a lack of maintenance of relevant information, lack of records, lack of staff training, and, consequently, lack of internal controls to prevent and to report suspicious transactions. For example, there was nondisclosure of transactions amounting to over US$ 10,000 made in 1 day by customers, albeit in different operations. The Tonkawa tribe was punished with a fine of US$ 1,000,000 and Edward E. Street received a fine of US$ 1,500,000, in accordance with an agreement with FinCEN, dated March 24, 2006. The Tonkawa tribe closed the Tonkawa Bingo and Casino as a result of this case.

4.5 Measures for Crime Prevention

It is not acceptable to possess a seemingly robust and aggressive system against money laundering when a sector that is completely vulnerable to all sorts of criminal practices still exists. Hence, a justified reflection must be taken from the moment the illicit funds migrate to "untouchable" sectors. To avoid the use of gambling as a means for money laundering, tools must be established to allow proper regulation of gambling houses, including the imposition of sanctions when they are negligent in preventing money laundering. It is important to monitor their activities through the implementation of the CDD, which monitors the performance of customers following international standards (i.e., a US$ 10,000 limit) as a framework for a more detailed evaluation. This limitation should be considered, regardless of the type of transaction made.

International cooperation between regulators is extremely important because it allows for the exchange of relevant information regarding gambling houses and junket activities. The exchange of information about experiences involving money laundering has proven to be a valuable way to detect, prevent, or counteract money laundering. Thus, we need to use certain mechanisms to exchange such information, including the following: (1) the adoption of an internal and permanent monitoring of customer activities, regardless of who they are; (2) constant internal and external training of employees; (3) designation of an employee or a group of employees to be in charge of monitoring the day-to-day operations of these gambling houses; (4) providing detailed information when demanded (such as name, address, identity, and activity), even when systems are used for automatic data; and (5) submission to civil punishment, irrespective of the criminal, through an administrative procedure managed by the FIU in the event of a breach of the duty to monitor as established by law and regulatory norms.

It is also vital to require special cards that are nontransferable and customized for each client, which would facilitate combating criminal activity within casinos and gambling houses. The presentation of such cards for identification to transact in this sector would be an especially helpful measure.

In addition, it is very important to maintain records for each transaction and customer identification, at least after a certain limit (US$ 3,000). It is necessary for customers to complete a particular transaction report that includes all operations where an exchange of cash for chips, gains, or games played occurs. A copy of this report should be delivered to the Internal Revenue Service (IRS), and winnings not listed in the report should not be delivered.

Further discussion is necessary regarding the prohibition of transactions when they involve the exchange of cash for cash, check, or wire transfer, and the issue of checks on behalf of third parties or delivering winnings to these third parties. Gambling houses have already given much attention to security issues, transparency, and marketing, with a focus on the financial transactions of customers and their behavior during games, in order to garner greater participation from customers. Similarly, there needs to be greater focus on these games and transactions in order to look for possible indicators of suspicious transactions, rather than simply collecting customer data. For example, if a client sends large sums of money for a casino through a bank and then, after a little gambling activity, requests that the balance be forwarded to an account in a tax haven, this should be reported as a suspicious activity transaction. The same should be required in the case of systematic requests for conversions of winnings into jewels, airline tickets, and tour packages.

In addition to the attention given to the physical movement of money, the gambling houses' client accounts must also be submitted to anti-money laundering

(AML) controls. Moreover, these proposed measures should be implemented within the regional offices and subsidiaries of gambling houses and casinos, even though these regional offices and subsidiaries are occasionally located in other countries. However, the provisions for facilities like safes and deposit services should remain covered by the official controls.

In addition to monitoring gambling houses, it is also imperative to have clear rules for junket activities. The organizers, promoters, and junket agents should be subject to special approval and specific rules regarding transparency in order to inhibit illicit activity. This must apply in the country where the gambling house is established as well as the country where the junket activity takes place. In addition to general junket supervision, high-roller customers who have special rooms and/or get special treatment deserve particular attention. Often, people with large amounts of money are considered special customers, but the managers of gambling houses must avoid making their business operate as a free zone for the commission of crimes by these special customers. Since the traditional high-roller environment tends to enable the commission of certain crimes, a policy that rewards diligent managers would be suitable for changing the existing culture. FIUs should also be allowed timely and proper access to information about these special customers. Furthermore, the regulation of these activities cannot be limited to foreign customers, nor should they be considered less important when citizens are forbidden from attending gambling houses.

Other options for preventing and detecting money laundering include the following:

• Anti-money laundering controls should be imposed on cruises, even though their activities are not limited to gambling, to ensure that they are obeying the rules under the jurisdiction of both the vessel's original country and the country where it is moored.
• Employees should be continuously trained so that they are able to effectively implement rules that combat money laundering in coordination with the traditional authorities on crime prevention who are in charge of regulating the games as a whole.
• Gambling houses should be subject to special audits focused on the prevention of money laundering. These audits can be performed by the regulated sector, as long as they are performed in harmony with the official agencies of crime prevention. Cooperation between the FIU and the regulatory sector is vital to the effectiveness of this process.
• Gambling houses should be required to report annual gross revenues greater than a certain value (e.g., US$ 1,000,000), as well as individual transactions with customers and total transactions per day that are greater than US$ 10,000.
• In jurisdictions with several gambling houses, police stations specialized in money laundering should be created in order to facilitate the collection of intelligence information related to the gambling houses and their customers.

- Gambling houses should be allowed to ban customers after considering their criminal history or other intelligence information. Such a policy cannot be considered illegal or abusive, since it is reasonably related to legitimate interests.
- Language should be added to any Money Laundering Act that requires suspicious activity reporting that goes beyond the identification of clients and their administrative responsibility, to include individuals or entities acting on promotion, brokerage, marketing, agency, management, or negotiation of gaming activity, including online or junket services.[20]

To summarize, the system of money laundering prevention should operate like a well-calibrated machine by making it mandatory for gambling houses to confidentially report suspicious transactions to the FIUs as well as provide a special client list. Failure to comply with these legal requirements, or unsatisfactory compliance, should lead to sanctions.

In the case of lotteries, important and simple measures would compel winners to register with their IRS number and provide a written justification.

Finally, control agencies, such as FIUs, the IRS, and the managers of lotteries, should create a national register of players divided by such factors as age group, typical bets, and betting location. The joint action of these entities would strengthen the supervision over people involved in gambling activities, making it easier to identify suspicious behaviors, report suspicious transactions, and incorporate policies aimed at detecting these economic crimes.

4.6   Enforcement Agencies

The government must be able to identify, disrupt, and dismantle networks that engage in money laundering and terrorism financing. These crimes have gained a degree of effectiveness and sophistication, but the FATF has compelled states to fight against the networks by using various fronts, including an aggressive policy that enables their discovery and eradication. Efforts have also been taken to track the data in the form of number of cases, convictions, arrests, seizures, confiscations, and effective regulatory actions.

The fight against money laundering must incorporate the work of all enforcement agencies to give a clear overview of the system, with a special focus on the environments that are most susceptible to money laundering. In addition to the traditional police officers and prosecutors, this task must be carried out by the Treasury Department, the Federal Reserve, the IRS, the social security agency, the State Department, the FIUs, and regulators of any activities related to gambling and sports,

---

[20] Within the sports sector, the language in the law on money laundering (Brazilian Law no. 9.613 of March 3, 1998, as amended by Law no. 12.683 of July 9, 2012) was especially valuable because it required communication in the following manner: "XV—individuals or companies that operate in the promotion, brokerage, marketing, brokering or negotiating rights transfer athletes, artists or fairs, exhibitions or similar events."

among others. Each enforcement agency should have a leadership role with respect to a particular category of conduct where they have specialized knowledge in order to detect illegal threats or actions that indicate money laundering.

It is not enough for these entities to act independently. Hence, collaboration among them and aggregated performances are essential components to combat crime. This collaboration is much more effective than independent investigation, since it allows for the exchange and sharing of data, experience, and intelligence information. Given the globalized nature of financial networks, it is not conducive for a country to take all necessary measures to combat financial crime when other countries do not impose similar measures. Therefore, the regulatory authorities and enforcement agencies should participate in AML domestic policy, but they should do so in conjunction with their international counterparts through bilateral or multilateral channels that enable coordination and mutual assistance.

The FIUs should continue to overcome the challenges posed by these crimes, but not solely through the supervision over traditional financial institutions. The requirement to monitor customers has to exceed the already-imposed limits to go beyond activities within financial institutions in order to enable access to information also relevant in gambling houses or lotteries. Thus, FIUs should return to engaging in advanced data analysis, with increased efforts in completing a regionalized assessment, if appropriate, of the threats in the gambling sector, including lotteries. This would provide an initial picture of the categorized information in order to detect potential vulnerabilities in existing controls.

In addition, the exchange of information with other FIUs of various countries is useful to obtain data on activities of money changers or money services businesses (MSBs). Moreover, the activities surrounding gambling houses should also be of concern to the IRS, compelling it to develop an aggressive campaign to reach a larger universe of activity subject to AML policy. Thus, the FIUs, in conjunction with the IRS, should act to strengthen the supervision of the people involved in gambling activities in order to identify suspicious behaviors and report suspicious transactions.

Many criminals, fearing the reporting activity of official financial institutions that inhibit the mobility of illicit funds, may turn to clandestine shipments as an alternative. While such criminals often structure transactions to bypass financial institutions and avoid potential reporting of suspicious activity, money launderers use these institutions to their benefit. In the USA, for example, a number of people acquired chips with illicit money in amounts below the limit where mandatory reporting would be imposed. Afterward, these people gave their chips to an individual who exchanged them for cash or check. The purpose of this process was to avoid reporting to the FIU. Twelve months later, it was discovered that the individual had been repaid US$ 1.1 million, but that at no time was this reported to the FIU.

Given these facts, it is important to structure enforcement agencies to detect money laundering and, if necessary, to draft legislative changes that help resolve the vulnerabilities found through investigations by these enforcement agencies.

Bibliography

ANISTIA poderá coibir o 'esquema Daslu'. Colecionadores afirmam que sonegação continuará sendo praxe no meio enquanto carga tributária não diminuir. *Folha de São Paulo*, Jan. 1, 2012, Ilustrada supplement, p. E5.

ANNUAL REPORT 2010–2011. Financial Action Task Force/Groupe d'Action Financière. fatf-gafi.org. Accessed Mar. 20, 2012.

APG reports deficiencies in Nepal's anti-money laundering compliance. Ekantipur.com. Aug. 20, 2011. Kantipur Publications Pvt.

ASCENSÃO, José de Oliveira. Branqueamento de Capitais: reacção criminal. Estudos de direito bancário. Coimbra: Coimbra, 1999.

ASSIS TOLEDO, Francisco de. *Princípios Básicos de Direito Penal—De acordo com a Lei n. 7.209, de 11.07.1984, e a Constituição de 1988*. São Paulo: Saraiva, 4th ed., 1991.

BALTAZAR JÚNIOR, José Paulo. *Crimes Federais*. Porto Alegre: Livraria do Advogado, 7th ed., 2011.

_____. *Crime Organizado e Proibição de Insuficiência*. Porto Alegre: Livraria do Advogado, 2010.

_____. *et alii. Lavagem de dinheiro. Comentários à lei pelos juízes das varas especializadas em homenagem ao Ministro Gilson Dipp*. Porto Alegre: Livraria do Advogado, 2007.

BARBOSA, Renato Rodrigues. Perito propõe estratégias de inteligência financeira no CJF. *Revista Perícia Federal*, Brasília, ano 5, no. 19, Nov/Dec 2004.

BARROS, Marco Antônio de. *Lavagem de capitais e obrigações civis correlatas: com comentários, artigo por artigo, à Lei 9.613/98*. São Paulo: Ed. Revista dos Tribunais, 2004.

_____. *Lavagem de dinheiro: implicações penais, processuais e administrativas. Análise sistemática da Lei n.º 9.613, de 3-3-1998*. São Paulo: Oliveira Mendes, 1998.

BARROSO, Luís Roberto. Da constitucionalidade do Projeto de Lei 3.115/97. *Revista de Direito Bancário, do Mercado de Capitais e da Arbitragem*, no. 16, pp. 199–210, Apr/June 2002.

BETTI, Francisco de Assis. *Aspectos dos crimes contra o sistema financeiro no Brasil*—comentários às Leis 7.492/86 e 9.613/98. Belo Horizonte: Del Rey, 2000.

BETTIOL, Giuseppe. *Direito penal*. Tradução brasileira e notas de Paulo José da Costa Júnior e de Alberto Silva Franco. São Paulo: Ed. Revista dos Tribunais, 1976.

BINDING, Karl. *La culpabilidad en derecho penal*. Transl. Manuel Cancio Meliá. Montevidéu, Buenos Aires: B de F, 2009.

BITENCOURT, Cezar Roberto. *Tratado de direito penal—Parte geral*. Vol.1, 17th ed. São Paulo: Saraiva, 2012.

_____. Crimes contra o sistema financeiro nacional praticados por administradoras de consórcios. Responsabilidade penal da pessoa jurídica. Atipicidade. *Revista dos Tribunais*, São Paulo, no. 735, Jan 1997.

_____. *Manual de direito penal*: parte geral. Vol. 1, 6th ed. São Paulo: Saraiva, 2000.

_____. Responsabilidade penal da pessoa jurídica à luz da Constituição Federal. *Boletim do IBC-Crim*, São Paulo, ed. esp. no. 65, April 1998.

BOLDT, Raphael. Delação premiada: o dilema ético. *Direito Net*. www.direitonet.com.br. Accessed Sept. 15, 2005.

BOULOC, Bernard. Coactivité en matière de publicité trompeuse. *Revue de Science Criminelle et de Droit Pénal Comparé*, [S.l.], p. 95, Jan/Mar 1995.

_____; STEFANI, Gaston; LEVASSEUR, Georges. *Droit pénal géneral*. 17th ed. Paris: Dalloz, 2000.

BROYER, Philippe et al. *La nouvelle économie criminelle*: criminalité financière—comment le blanchiment de l'argent sale et le financement du terrorisme sont devenus une menace pour les entreprises et les marchés financiers. Paris: Éditions d'Organisation, 2002.

BRUNO, Aníbal. *Direito penal*. Rio de Janeiro: Forense, 2009.

CAEIRO, Pedro. *Branqueamento de capitais*. Manual distributed in coursework offered 17–21 October 2005 and sponsored by the OAS and the Justice Ministry for Brazilian judges and prosecutors.

CAPEZ, Fernando. *Curso de direito penal—Parte geral*. Vol. 1, 16th ed. São Paulo: Saraiva, 2012.

CASSELLA, Stefan D. The Money Laundering Statutes (18 U.S.C. §§ 1956 and 1957). *United States Attorney's Bulletin*. Washington, DC, Vol. 55, No. 5, Sept. 2007.

CEREZO MIR, José. *Curso de derecho penal español—Parte general*. Vols. II & III. Madrid: Tecnos, 2001.

CERVINI, Raúl. Macrocriminalidad económica. *Revista Brasileira de Ciências Criminais*. Vol. 11, pp. 50–79. São Paulo: Ed. RT, July-Sept 1995.

_____ et al. *Lei de lavagem de capitais*. São Paulo: Ed. RT, 1998.

CHAIRMAN'S SUMMARY, Paris Plenary, June 20–23, 2006. Financial Action Task Force/ Groupe d'Action Financière. Fatf—Gafi. http://www.fatf-gafi.org. Accessed July 3, 2006.

CONSELHO DA JUSTIÇA FEDERAL. Centro de Estudos Judiciários, Secretaria de Pesquisa e Informação Jurídicas. *Uma análise crítica da lei dos crimes de lavagem de dinheiro*. Brasília: CJF, 2002.

CONTE, Philippe; LARGUIER, Jean. Le recel de choses et le blanchiment. *Droit pénal des affaires*. Paris: Dalloz; Armand Colin, 2004.

_____; _____. *Droit pénal des affaires*. 11th ed. Paris: Paris: Dalloz; Armand Colin, 2004.

CORRÊA, Hudson. Juiz denuncia uso de gado para lavagem de dinheiro. *Folha de São Paulo*, São Paulo, July 24, 2005. Dinheiro, B7.

COSTA, José de Faria. O branqueamento de capitais (algumas reflexões à luz do direito penal e da política criminal). In CORREIA, Eduardo et al. *Direito penal econômico e europeu:* textos doutrinários. Vol. 2. Coimbra: Coimbra Ed., 1999.

_____. *Breves reflexões sobre o Decreto-Lei n. 207-B/75 e o direito penal econômico.*

COUNCIL FOR FINANCIAL ACTIVITIES CONTROL — COAF. https://www.coaf.fazenda. gov.br/conteudo/estatisticas/comunicacoes-recebidas-por-segmento/. Accessed May 18, 2012.

CROCQ, Jean-Christophe. *Le guide des infractions*. 3rd ed. Paris: Dalloz, 2001.

CUESTA AGUADO, Paz M. de la. *Causalidad de los delitos contra el medio ambiente*. Valencia: Tirant lo Blanch, 1995.

D'AMORIM, Sheila, and SOUZA, Leonardo. BC falha no combate à lavagem, afirma TCU. *Folha de São Paulo*. São Paulo, Oct. 13, 2005. Caderno Dinheiro, B3.

DE CARLI, Carla Veríssimo (Org.). *Lavagem de dinheiro: prevenção e controle penal*. Porto Alegre: Verbo Jurídico, 2011.

DELMANTO JÚNIOR, Roberto. Justiça especializada para os crimes de lavagem de dinheiro e contra o Sistema Financeiro Nacional (a inconstitucionalidade da Resolução n° 314, de 12/5/2003, do Conselho da Justiça Federal). *Revista do Advogado*, São Paulo, ano 24, no. 78, pp. 95–102, Sept. 2004.

DELMAS-MARTY, Mireille. *Droit pénal des affaires*. 3rd ed. Paris: Presses Universitaire de France, 1990.

_____, and GIUDICELLI-DELAGE, Geneviève. *Droit pénal des affaires*. 4th ed. Paris: Presses Universitaire de France, 2000.

DE SANCTIS, Fausto Martin. *Lavagem de Dinheiro: Jogos de Azar e Futebol. Análise e Proposições*. Curitiba: Juruá, 2010.

_____. *Crime Organizado e Lavagem de Dinheiro. Destinação de Bens Apreendidos, Delação Premiada e Responsabilidade Social*. São Paulo: Saraiva, 2009.

DOTTI, René Ariel. *Curso de direito penal—Parte geral*. 2nd ed. Rio de Janeiro: Forense, 2004.

FARIA, Bento de. *Código Penal brasileiro* (comentado). Vol. 5, 2nd ed. Rio de Janeiro: Record, 1959.

FINANCIAL ACTION TASK FORCE—FATF Money Laundering through the Football Sector Report. Available at http://www.fatf-gafi.org/topics/methodsandtrends/documents/moneylaunderingthroughthefootballsector.html. Last updated Feb. 1, 2012. Accessed on May 10, 2013.

_____. Vulnerabilities of Casinos and Gaming Sector, March 2009. In: http://www.fatf-gafi.org/ mwg-internal/de5fs23hu73ds/progress?id=g0bbqv2M5r. Accessed on June 10, 2013.

FRAGOSO, Heleno Cláudio. *Lições de direito penal*: a nova parte geral. 8th ed. Rio de Janeiro: Forense, 1985.

FREIRE JÚNIOR, Américo Bede and SENNA MIRANDA, Gustavo. *Princípios do processo penal—Entre o garantismo e a efetividade da sanção*. São Paulo, Ed. RT, 2009.

GÁLVEZ VILLEGAS, Tomás Aladino. Delito de lavado de activos: análisis de los Tipos Penales Previstos en la Ley n° 27.765. *Magazine of the Office of the Public Prosecutor of Peru*, ("Fiscalía de la Nación"), pp. 46–62, Feb 2005.

GAMA, Guilherme Calmon Nogueira da; GOMES, Abel Fernandes. *Temas de direito penal e processo penal*: em especial na Justiça Federal. Rio de Janeiro: Renovar, 1999.

GARCIA, Basileu. *Instituições de direito penal*. Vol. 1. 4th ed. São Paulo: Max Limonad, 1978.

GARCIA, José Ángel Brandariz. *El delito de defraudación a la seguridad social*. Valencia: Tirand lo Blanch, 2000.

GARCIA, Plínio Gustavo Prado. Arbítrio e inconstitucionalidade na lei de lavagem de dinheiro ou bens. *Informativo do I.A.S.P.*, São Paulo, no. 45, Mar/Apr 2000.

GODINHO, Jorge Alexandre Fernandes. *Do crime de "branqueamento" de capitais*: introdução e tipicidade. Coimbra: Almedina Ed., 2001.

GODOY, Arnaldo Sampaio de Moraes. *Direito tributário comparado e tratados internacionais fiscais*. Porto Alegre: Sergio Antonio Fabris, 2005.

_____. *Direito tributário nos Estados Unidos*. São Paulo: Lex, 2004.

GODOY, Luiz Roberto Ungaretti de. *Crime organizado e seu tratamento jurídico penal*. Rio de Janeiro: Elsevier, 2011.

GOMES, Luiz Flávio. *Direito penal—Parte geral*. 2nd ed. São Paulo: Ed. RT, 2009.

_____. Crime organizado: que se entende por isso depois da Lei n° 10.217/01?—Apontamentos sobre a perda de eficácia de grande parte da Lei 9.034/95. *Jus Navegandi*. www1.jus.com.br. Accessed June 1, 2004.

GLENNY, Misha. *McMafia*: Crime sem fronteiras. Trad. Lucia Boldrini. São Paulo: Companhia das Letras, 2008.

GUIBU, Fábio. Para ministro, há cultura de crimes. Folha de São Paulo. São Paulo, Oct. 28, 2012. Caderno Brasil, A7.

HAMPTON, Alan. *Sources of Information in a Financial Investigation*. United States Attorney's Bulletin. Washington, DC, Vol. 55, No. 5, Sept. 2007.

HUNGRIA, Nélson. *Comentários ao Código Penal—Arts. 1 a 10, 11 a 27, 75 a 101*. 4th ed. Rio de Janeiro: Forense, 1958.

INFORME ANNUAL 2011. *Unidad de Información Financiera*. Buenos Aires: Departamento de Prensa, Ministerio de Justicia y Derechos Humanos/Presidencia de la Nación, 2012.

INSUFFICIENT laws to tackle terror-funding: FATF. *Mint*. WLNR 14346920. July 18, 2010.

JEANDIDIER, Wilfried. *Droit Penal dês Affaires*. Paris: Dalloz, 1996.

JESUS, Damásio Evangelista de. *Direito penal—Parte geral*. Vol. 1, 33rd ed. São Paulo: Saraiva, 2012.

_____. *Imputação objetiva*. 3rd ed. São Paulo: Saraiva, 2007.

JOSEPH, Lester. *Suspicious Activity Reports Disclosure and Protection*. United States Attorney's Bulletin. Washington, DC, Vol. 55, No. 5, Sept. 2007.

_____; ROTH, John. *Criminal Prosecution of Banks Under the Bank Secrecy Act*. United States Attorney's Bulletin. Washington, DC, Vol. 55, No. 5, Sept. 2007.

JOY, Stephen; WARREN KINDT, John. Internet Gambling and the Destabilization of National and International Economies. Time for a Comprehensive ban on Gambling Over the World Wide Web. HeinOnline, citation:80 Denv. U. L. Rev. 111 2002–2003, p. 111.

LACERDA DA COSTA PINTO, Frederico de. Crimes econômicos e mercados financeiros. Revista Brasileira de Ciências Criminais, São Paulo, n. 39, pp. 28–62, July/Sept 2002.

LEITE, Carlos Eduardo Copetti. Força-tarefa: conceito, características e forma de atuação. *Revista dos Procuradores da República*, ano 1, no. 2, pp. 8–9, Sept 2004.

LYRA, Roberto. *Criminalidade econômico-financeira*. Rio de Janeiro: Forense, 1978.

MACHADO, Agapito. *Crimes do colarinho branco e contrabando/descaminho*. São Paulo: Malheiros Ed., 1998.

MACHADO, Miguel Pedrosa. A propósito da revisão do Decreto-Lei n.° 28/84, de 20 de janeiro (Infracções antieconômicas). In: CORREIA, Eduardo et al. *Direito penal econômico e europeu:* textos doutrinários. Vol. 1. Coimbra: Coimbra Ed. 1998.

MAIA, Rodolfo Tigre. *Dos crimes contra o sistema financeiro nacional:* anotações à Lei Federal n.° 7.492/81986. São Paulo: Malheiros Ed., 1996.

_____. *Lavagem de dinheiro—lavagem de ativos provenientes de crime. Anotações às disposições criminais da Lei n.° 9.613/1998.* São Paulo: Malheiros Ed., 1999.

MARQUES DA SILVA, Germano. Direito Penal Português—parte geral. Vol. II. Lisboa: Verbo, 1999.

MENDEZ RODRIGUEZ, Cristina. *Los delitos de peligro y sus técnicas de tipificación.* Madrid, 1993.

MENDRONI, Marcelo Batlouni. *Crime de lavagem de dinheiro.* São Paulo: Atlas, 2006.

_____. Crime de lavagem de dinheiro: consumação e tentativa. *Ultima Instância:* revista jurídica. www.ultimainstancia.com.br. Accessed June 21, 2012.

_____. Delação premiada. *Ultima Instância:* revista jurídica. www.ultimainstancia.com.br. Accessed 28 August 2005.

MESSINEO, Francesco. Máfia e Crime de Colarinho Branco: uma nova abordagem de análise. *Novas tendências da criminalidade transnacional mafiosa.* Organizers: Alesandra Dino and Wálter Fanganiello Maierovitch. São Paulo: Unesp Ed., 2010.

MIR, José Ricardo Sanchís and Vicente Garrido Genovés. *Delincuencia de cuello blanco.* Madrid: Instituto de Estudios de Política, 1987.

MIR PUIG, Santiago. *Derecho penal—Parte general (fundamentos y teoría del delito).* Barcelona: Promociones Publicaciones Universitarias, 1984.

_____. *Introducción a las bases del derecho penal.* Montevideo/Buenos Aires: B de F, 2003.

MORAES, Alexandre. Direito Constitucional. 6th ed. São Paulo: Atlas, 1999.

OLIVEIRA, Odilon. Lavar dinheiro com gado é muito fácil. *Jornal da Tarde,* June 25, 2007.

OLIVEIRA, William Terra de; CERVINI, Raúl; GOMES, Luiz Flávio. Terra de. *Lei de lavagem de capitais.* São Paulo: Ed. Revista dos Tribunais, 1998.

OVERVIEW of Gaming Worldwide, Casino City, Global Gaming Almanac, 2007, available at: http:/casinocitypress.com/GamingAlmanac/globalga-mingalmanac and Gaming Data Report: Global Internet Gambling Revenue Estimates and Projections, *Christiansen Capital Advisors,* 2005, http:/www.cca-i.com.

REALUYO, Celina B. *It's All about Money: Advancing Anti-Money Laundering Efforts in the U.S. and Mexico to Combat Transnational Organized Crime.* Washington, DC: Woodrow Wilson International Center for Scholars. Mexico Institute, May 2012.

RIDER, Barry. The Financial World at Risk: The Dangers of Organized Crime, Money Laundering and Corruption. *Managing Auditing Journal.* no. 7, 1993.

SANTIAGO, Rodrigo. O "branqueamento" de capitais e outros produtos do crime: contributos para o estudo do art. 23.° do Decreto-Lei n.° 15/93, de 22 de janeiro, e do regime de prevenção da utilização do sistema financeiro no "branqueamento" (Decreto-Lei n.° 313/93, de 15 de setembro). In: CORREIA, Eduardo et al. *Direito penal econômico e europeu:* textos doutrinários. Vol. 2. Coimbra: Coimbra Ed., 1999.

SANTOS, Cláudia Maria Cruz. *O crime de colarinho branco (da origem do conceito e sua relevância criminológica à questão da desigualdade na administração da justiça penal).* Coimbra: Coimbra Ed., 2001.

SILVA, Germano Marques da. *Direito penal português—Parte geral.* Vol. II. Lisbon: Verbo, 1999.

SIMÕES, Euclides Damaso. Manual distributed in coursework offered 17–21 October 2005 and sponsored by the OAS and the Justice Ministry for Brazilian judges and prosecutors.

SIMPÓSIO sobre direito dos valores mobiliários. Série cadernos do Centro de Estudos Judiciários do Conselho da Justiça Federal, nos. 15 and 16. Brasília: 1998 and 1999.

SUÁREZ-INCLÁN DUCASSI, María Rosa. Financial Regulations and Tax Incentives with the Aim to Stimulate the Protection and Preservation of Cultural Heritage in Spain. *Art and Cultural Heritage: Law, Policy, and Practice.* New York: Cambridge University Press, edited by Barbara T. Hoffman, 2006.

SUTHERLAND, Edwin H. *El Delito de Cuello Blanco—White Collar Crime—The Uncut Version.* Buenos Aires: Editorial IBdeF, 2009.

_____. *White-Collar Crime—The Uncut Version.* New Haven: Yale University Press, 1983.

THE SAR Activity Review, Trends, Tips & Issues. *Published under the auspices of the BSA Advisory Group,* Issue 12 FinCEN, Oct. 2007.

THOMPSON, Erin. The Relationship between Tax Deductions and the Market for Unprovenanced Antiquities. 33 Colum. J.L. & Arts. 241, 2010.

TOLEDO, Francisco de Assis. *Princípios básicos de direito penal.* São Paulo: Saraiva, 2000.

_____. *Princípios básicos de direito penal—De acordo com a Lei n. 7.209, de 11.7.1984, e a Constituição de 1988.* 4. ed. São Paulo: Saraiva, 1991.

TSITSOURA, Aglaia. Les travaux du Conseil de l'Europe. *Revue Internationale de Droit Pénal,* [S.l.], vol. 54, no. 3/4, 1983.

VERMELLE, Georges. *Le nouveau droit pénal.* Paris: Dalloz, 1994. (Série Connaissance du droit).

VIEIRA, Ariovaldo M. Coordenador. Temas relevantes no direito penal econômico e processual penal. São Paulo: Ed. Federal, 2007.

VILLA, John K. Banking Crimes: Fraud, Money Laundering and Embezzlement, Vol. 2, App. 2A5-6.

VULNERABILITIES of Casinos and Gaming Sector, March 2009. Available at http:/www.gafi-fatf.org. Accessed April 7, 2010.

WELZEL, Hans. *Derecho penal alemán: parte general.* 11th ed. (4th ed. Castellana). Transl. Juan Bustos Ramírez & Sérgio Yánez Pérez. Santiago: Jurídica, 1997.

ZAFFARONI, Eugenio Raúl. *Moderna dogmática del tipo penal.* Lima: Aras Editores, 2009.

_____; PIERANGELI, José Henrique. *Manual de direito penal brasileiro—Parte geral.* 2nd ed. São Paulo: Ed. RT, 2005.

_____; _____. *Tratado de derecho penal. Parte general.* Buenos Aires, 1998.

ZANCHETTI, Mário. Il ricciclaggio di denaro proveniente da reato. Milan: Giuffrè, 1997.

Chapter 5
Illegal Betting and Internet Gambling

Illegal betting is prevalent in the worlds of sport, gambling, and money laundering. Criminals use illegal betting as a means to launder their gains from criminal activities. Therefore, premiums arising from such bets (and here we could also talk about "investments") with dirty money are also illegal and subject to seizure, freezing, and confiscation.

The modus operandi of some of these money launderers is to acquire lottery tickets from the actual awarded people. Knowing awarded people or even lottery owners from where tickets are bought, criminals can apply for the reward of the winner. The scheme is possible only due to collusion between employees or lottery owners and offenders. Thus, the dirty money changes hands. Repeated lottery awards to the same individual in short periods of time challenge the most rudimentary notions of mathematical probability. Although the possibility of several awards to the same person is near zero, the principle of in dubio *pro reo* could always be invoked.

Lottery owners sell lottery tickets through enticing winners. That is why it is important for the government to control and regulate these lottery houses. In Brazil, the responsible government agency would be the Caixa Econômica Federal.

The Financial Action Task Force (FATF) reports that money launderers use football in the following typologies: (1) acquisitions and investments in football clubs, (2) international purchase and transfer of players, (3) acquisition and sale of game tickets, (4) bets, and (5) misuse of image rights, sponsorships, and advertising.

Hervé Martin Delpierre, who made a documentary titled "24h chrono," released on May 8, 2013, affirms that the morals of sport are now in doubt. He reveals that of the 15,000 sports betting websites ("paris sportifs")[1] in the world, 85% are illegal. These websites generate more money than sport itself; 10–100 times more in certain events. Illicit deals can take place not only about the outcome of a game but also the number of points or goals, the amount of corners or yellow or red cards, etc.

---

[1] Sixty percent of sports betting websites in the world are for football.

F. M. De Sanctis, *Football, Gambling, and Money Laundering*,
DOI 10.1007/978-3-319-05609-8_5, © Springer International Publishing Switzerland 2014

It is also possible to buy off players, clubs, and referees to fix games. Of the 27,000 football games played each year under the rules of the Union of European Football Associations (UEFA), 7 % are suspected of manipulation.[2]

Philippe Kern, founder and director of KEA European Affairs and director of the French think tank, Sport et Citoyenneté ("Sport and Citizenship"), says that a recent study called "Match-fixing in Sport," led by the European Commission, made recommendations, including criminal penalties, for European countries to take action.[3] The European Commission also published an extensive "White Paper on Sport," detailing money laundering and corruption in the sport sector.[4]

Players themselves, who have time and money in excess, also engage in Internet betting and, as a result, have been targeted for their bets and debts. Hervé Martin Delpierre reveals in his documentary, the story of a goalkeeper who ended up in the hands of the Camorra and later received additional threats.

To combat fraud, it is relevant to envision a way to control such gambling websites and the flow of money involved. The EC's recommendation to create a crime of "sporting fraud" can also be considered. It is necessary to sensitize players, making them able to resist any pressures and report them to authorities. The economic situation of sport should be sufficiently viable in a way that players are not tempted

---

[2] In Du blanchiment d'argent au trucage de matches. Paris en ligne: les mafias truquent le jeu. *Dernières Nouvelles D'Alsace (DNA)*. April 27, 2012, http://www.dna.fr/justice/2012/04/27/les-mafias-truquent-le-jeu.

[3] Philippe Kern, *The Fight against corruption in Sport is a major European issue*. EurActiv. com, October 5, 2012, http://www.euractiv.com/culture/fight-corruption-sport-major-eur-analysis-515227.

[4] The European Commission's "White Paper on Sport" states the following:

> Sport organizations are generally aware of these problems and have for some time been discussing them with governmental actors. The need for sport organizations to be transparent was recognized by participants at the conference "Rules of the Game", which took place in Brussels in 2001. In fact, it is one of the key aspects of the conference report. The problem has also been recognized in a number of reports produced by sport organizations, including the "Stevens Report" on Premier League Transfers.
>
> One of the reasons why the Independent European Sport Review was launched was that it identified "a range of problems—such as doping, corruption, racism, illegal gambling, money-laundering, and other activities detrimental to the sport—where only a holistic approach between football and the EU and national authorities will be truly effective." The Review put these problems on record and identified the following key problem areas: "player transfers, payments to agents, investment in clubs and a variety of other commercial deals associated with football, such as sponsorship."
>
> Corruption in the sport sector may frequently be a reality and, given the sector's high degree of internationalization, is often likely to have cross-border aspects. Corruption problems which have a European dimension need to be tackled at European level.
>
> Corruption is particularly damaging for sport as it raises a credibility problem for sport associations. The sport sector cannot tackle the problem alone. Many major sport organizations have come to realize that they need to work more closely with governmental actors, including law enforcement bodies.

European Commission, http://ec.europa.eu/sport/white-paper/swd-the-organisation-of-sport_en.htm, accessed May 7, 2013.

by organized crime. A mandatory central body of information with the necessary awareness of sports federations and confederations should be the first step to avoid illicit behaviors.

Gambling conducted via the Internet has drawn a lot of attention. In Brazil, casinos and bingos are considered illegal.[5] Ian Abovitz reveals that "[i]n the United States, courts have traditionally recognized gambling as an area reserved for state regulation pursuant to the Tenth Amendment of the U.S. Constitution. Currently, all 50 states and the District of Columbia conduct some form of gambling regulation, ranging from full legalization in Nevada to blanket prohibition in Hawaii and Utah."[6]

Usually, these regulations are designed to generate tax revenue while also providing for the safety of players and operators by limiting the social concerns associated with gambling.

John Warren Kindt and Stephen W. Joy state that "[a] majority of the money generated by Internet casinos went untaxed, created more untaxable money flow, and reduced taxable economic activities."[7]

Although the traditional methods of regulating gambling have been effective, application to the Internet has proven difficult as the boundless nature and wide accessibility of the medium are widely believed to intensify social concerns.[8]

On October 13, 2006, President George W. Bush signed into law the Unlawful Internet Gambling Enforcement Act of 2006 (UIGEA),[9] which prohibits the acceptance of payment of wagers by financial institutions. The UIGEA bans Internet gambling by forcing financial institutions to prevent financial payments of wagers from bank accounts and other financial instruments.[10]

In March 2011, the European Commission opened a public consultation called "On-line gambling in the Internal Market." Its main aim was to identify public policy challenges and market issues referring to the differing regulatory models of national authorities in the European Union (EU).[11] The consultation welcomed

---

[5] Law 8.672 of June 7, 1993, called the Zico Act, gave clubs the choice to become companies. In turn, Law 9.615 of 24.03.1998, called Pele Act, revoked Zico Act and was later amended by Law 9.981 of July 14, 2000, called the Maguito Vilela Act, which revoked, in art. 2, the chapter devoted to bingo, thus outlawing bingo completely. Law 10.671 of May 15, 2003, addressed the financial transparency of management, established offenses, and considered sport as a cultural expression of the country.

[6] Ian Abovitz, *Why The United States Should Rethink Its Legal Approach To Internet Gambling: A Comparative Analysis Of Regulatory Models That Have Been Successfully Implemented In Foreign Jurisdictions.* 22 Temp. Int'l & Comp. L. J. 437 (2008), p. 438.

[7] John Warren Kindt and Stephen W. Joy, *Internet Gambling and the Destabilization of National and International Economies: Time for a Comprehensive Ban on Gambling over the World Wide Web.* 80 Denv. U. L. Rev. 111 (2002).

[8] *Id.*

[9] 31 U.S.C. §§ 5361–5367 (Supp. 2007).

[10] *See* Michael Blankenship, *The Unlawful Internet Gambling Enforcement Act: A Bad Gambling Act? You Betcha!,* 60 Rutgers L. Rev. 485 (2008).

[11] The Consultation took place March 24–July 31, 2011. *See* http://ec.europa.eu/internal_market/consultations/2011/online_gambling_en.htm, accessed June 7, 2013.

views of interested stakeholders, including citizens and private and public gambling operators. They were invited to share expertise and contribute data.

According to the "On-line gambling in the Internal Market" from the European Commission, online gambling services are today widely available and used in the EU. The economic significance of the sector is growing at a very high speed. The advent of the Internet and the growth of online gambling opportunities are posing regulatory challenges as these forms of gambling services are subject to national regulatory frameworks that vary rather significantly between member states. These frameworks can be broadly categorized into either licensed operators operating within a strictly regulated framework or strictly controlled monopolies (state-owned or otherwise).

A number of member states have also embarked on a review of their gambling legislation to account for these new forms of service delivery. Furthermore, the growth of online gambling opportunities has given rise to the growth of an unauthorized market, which consists of unlicensed illegal gambling and betting activity, including from third countries and operators licensed in one or more member states offering gambling services in other member states without having obtained the specific authorization in those countries.

Online gambling has recently generated an overwhelming interest worldwide. For Anurag Bana, it "is being seen as a potential trading sector that could assist countries with a 'booster shot' in reducing their accumulated fiscal deficits in an effort to encourage domestic and world economies."[12]

These developments pose regulatory and technical challenges, and they also give rise to societal and public-order issues, such as the protection of consumers from fraud and the prevention of gambling addiction.

As online gambling is a global phenomenon, an effective international regulator should be formed to monitor the management, accountability, efficiency, and sector proportionality of the stakeholders involved in order to sustain market confidence in trading by promoting public understanding in addition to maintaining an appropriate degree of protection for consumers.

According to Jon Mills, "while facilitating commerce and communication, the Internet also facilitates the ability of criminals to elude the laws of any, and every, nation."[13]

The Internet provides individuals worldwide with the ability to communicate and exchange information across national boundaries and continents. The project to connect scientists and defense agencies has united the globe with access to information, available anywhere, at any time. And it has also connected criminals and people with criminal purposes.

---

[12] Anurag Bana, *Online Gambling: An Appreciation of Legal Issues.* 12 Bus. L. Int'l 335 (2011), p. 335.

[13] Jon Mills, *Internet Casinos: A Sure Bet for Money Laundering.* 19 Dick. J. Int'l L. 77 (2000), p. 83.

Bibliography

ABOVITZ, Ian. *Why The United States Should Rethink Its Legal Approach To Internet Gambling: A Comparative Analysis Of Regulatory Models That Have Been Successfully Implemented In Foreign Jurisdictions.* 22 Temp. Int'l & Comp. L. J. 437 (2008).

BANA, Anurag. *Online Gambling: An Appreciation of Legal Issues.* 12 Bus. L. Int'l 335, 2011.

BLANKENSHIP, Michael. *The Unlawful Internet Gambling Enforcement Act: A Bad Gambling Act? You Betcha!.* 60 Rutgers L. Rev. 485, 2008.

CONSULTATION on online gambling in the Internal Market. The European Commission. http://ec.europa.eu/internal_market/consultations/2011/online_gambling_en.htm, accessed June 7, 2013.

DELPIERRE, Hervé Martin. Du blanchiment d'argent au trucage de matches. Paris en ligne: les mafias truquent le jeu. *Dernières Nouvelles D'Alsace (DNA).* April 27, 2012, http://www.dna.fr/justice/2012/04/27/les-mafias-truquent-le-jeu.

JOY, Stephen; KINDT, John Warren. *Internet Gambling and the Destabilization of National and International Economies: Time for a Comprehensive Ban on Gambling over the World Wide Web.* 80 Denv. U. L. Rev. 111 (2002).

KERN, Philippe. *The Fight against corruption in Sport is a major European issue.* EurActiv.com, October 5, 2012, http://www.euractiv.com/culture/fight-corruption-sport-major-eur-analysis-515227.

LE Sport en danger: corruption, paris illégaux, blanchiment d'argent, agents des sportifs et des clubs, arbitres etc…, Le portail de référence pour l'Espace de Liberté, Sécurité et Justice, March 4, 2011, http://europe-liberte-securite-justice.org/2011/03/04/le-sport-en-danger-corruption-paris-illegaux-blanchiment-d%E2%80%99argent-agents-des-sportifs-et-des-clubs-arbitres-etc-%E2%80%A6/.

MILLS, Jon. *Internet Casinos: A Sure Bet for Money Laundering,* 19 Dick. J. Int'l L. 77, 2000.

WHITE Paper on Sport. European Commission, http://ec.europa.eu/sport/white-paper/swd-the-organisation-of-sport_en.htm, accessed May 7, 2013.

Chapter 6
The Use of Illegal and Disguised Instruments for Payments (Cash/E-money, Offshore Accounts, NGOs) by Organized Crime

The inappropriate use of wire transfers, prepaid cards, moneychangers, digital currencies, and stored value instruments, as well as NGOs, trusts, associations, and foundations, facilitate criminal activity. Thus, more adequate and coordinated ways to prevent money laundering must be studied.

When people or companies seek to send or receive money from unlawful behavior across national borders, undetected by government institutions, they have come to rely more and more on untraceable transfers. Examples include cash, e-money, dollar or Euro wire transfers (operated by agents known as dollar-changers (*doleiros*) whose activities stretch the legal envelope), use of offshore accounts, and NGOs. Money laundering can also occur through fraudulent payments when, for example, part of the amount payable is left out and paid instead in cash or some untraceable means and delivered to the money launderer under the table.

Organized crime has defied government control because one of the most effective instruments for crime fighting is to cut off its financing and confiscate the criminal proceeds, but this is impossible when large money transfers are unknown to and undetectable by authorities.

The following cases provide illustrative examples of the use of fraudulent payments:

(a) **Case I:** Two South Americans were arrested, one carrying US$ 7,000.00 and the other carrying US$ 6,182.00. One of the arrested persons argued that he was the lawyer of a goalkeeper of a big Brazilian football club, and that the money seized from them (around US$ 14,000.00) was the lawyer's fees from a negotiation of that football player. They were charged with attempt to commit a financial crime (capital flight). Once the settlement conditions were fulfilled in mid-2007, the culpability was extinguished and the proceeding ended.

(b) **Case II:** An Eastern European tried to board a plane to Western Europe, taking US$ 24,500.00 under his clothes. He had come to Brazil to select football players to be bought by the football club of another Eastern European country. He stated

---

For more information about illegal payments, see, DE SANCTIS, Fausto Martin. *Money Laundering Through Art. A Criminal Judicial Perspective.* Heidelberg: Springer International Publishing Switzerland, 2013.

F. M. De Sanctis, *Football, Gambling, and Money Laundering,*
DOI 10.1007/978-3-319-05609-8_6, © Springer International Publishing Switzerland 2014

that he had bought two football players from a Brazilian club for US$ 120,000.00 and received US$ 20,000.00 commission. The proceeding was suspended after 2 years, due to a settlement.

(c) **Case III**: Persons allegedly related to Investment Company "I" and to popular Brazilian Soccer Team "X" were charged (including the former president and former vice-president of Team "X"). According to the prosecution, a Western European multimillionaire ("A"), in a period of just 10 years, turned from an obscure and poorly paid professor to an influential, powerful, and multimillionaire politician. According to public documents received from the prosecutor's office in that country, "A" was under several police investigations for various crimes, including: (a) public capital diversion, interests, and contract fines, and money laundering committed by a criminal organization ("A" fled before the proceeding started); (b) controlling an organized group and using money diverted from fake commercial activities to buy real estate, including a cottage for the daughter of "A"; (c) embezzlement committed by a criminal organization; and (d) other crimes. A Middle-Eastern citizen, "C" introduced himself as the representative of Investment Company "I," an offshore company headquartered in Western Europe. At the time, the company did not even exist, in fact or formally, hence it did not even have any background that could give its credibility. "C" was unknown in the business world until some years before when, along with another fellow citizen from the Middle East, he set up and headquartered an investment fund in a Caribbean tax haven, and acquired 85 % a big publishing group. Sometime later, "A" "bought" the shares of his front man "C." The owners of Investment Company "I"—the president and the vice-president of Team "X" and "C" (the front man of "A")—signed a contract granting the licensing of intellectual property to "I," which would have the right to 51 % of the net profit made by Team "X." In compensation, "I" was obliged to invest US$ 35 million in "X." Wire taps conducted by the Federal Police showed that the suspects, in several opportunities, referred to "A" as the man who had the power of decision. In the partnership contract, "I" had to organize a limited company in Brazil and had to pay its capital stock in reals (Brazilian currency) for an amount equivalent to US$ 20 million, part of the US$ 35 million initially set forth in the pre-contract. To create such limited company, a Brazilian Law Firm "L" was hired and acted as partner to guarantee money inflows from abroad. Afterward, the lawyers sold the shares to the three offshore companies. The first two were headquartered in the Caribbean tax haven, and the third was in Western Europe, at an address that was subsequently confirmed to have just been used by a fitness center that belonged to "C" (the front man of "A"). One world famous football player was purchased for more than US$ 20 million and another one for more than € 8 million, which were paid abroad and not with the money brought to Brazil. The dates, amounts, and accounts were not revealed. Charges were finally brought for conspiracy, association in a criminal organization, and money laundering, among other offenses. The Brazilian criminal court's decision accepting the charges was based on the UN Convention against Corruption, supported by documents from Eastern and Western European countries, which listed evidence primarily obtained through wire taps. An order was issued to freeze the money deposited in the accounts, funds, and any kind of investments kept by "I" and of the money that might have been credited to accounts held by "X" due to possible exchange contracts with the mentioned offshore companies. It was also ordered that "X" provide the list of all players purchased with money from its partnership with "I." The trial had to start anew because of a decision by the Supreme Court of Brazil, and there is no final decision yet.[1]

In conclusion, in the first two cases above, involving persons arrested for trying to leave the country with an amount of cash in excess of the allowed limit, it is evident that the money seized was the proceeds of a suspicious negotiation with professional athletes because the negotiation itself was done with cash. In the third case, the football club investors accused of money laundering used a front person, shell companies, a tax haven, and specialized law firms.

A tax haven is a country that tries to attract nonresident funds by offering light regulation, low or zero taxation, and secrecy. Tax havens facilitate illegal activities, including money laundering and outright tax evasion. There are around 50–60 such havens in the world and nobody really knows how much money is stashed away: estimates vary from below to way above US$ 20 trillion.

The best way to combat illegal activities is transparency, which allows for more information to be collected and shared by law enforcement. It is important to crack down on the use of nominal shareholders and directors that hide the provenance of money.[2]

The Financial Action Task Force (FATF, or *Groupe d'Action Financière sur le blanchiment des capitaux*, GAFI), as stated before, established 40 Recommendations that highlight the importance of transparency to prevent money laundering. It recommends that participating nations obtain detailed information on all parties to wire transfers, cash, and border transactions, from both senders and beneficiaries, for monitoring purposes. Countries should establish policies to supervise and monitor nonprofit organizations, so as to obtain real-time information on their activities, size, and other important features such as transparency, integrity, and best practices (Recommendation No. 8); financial institution secrecy laws, or professional privi-

---

[1] No. 2006.61.81.008647-8, 6th Federal Trial Court in São Paulo/Brazil, specialized on money laundering.

[2] According to a 2013 article in *The Economist*, "Not all these havens are in sunny climes; indeed not all are technically offshore. Mr Obama likes to cite Ugland House, a building in the Cayman Islands that is officially home to 18,000 companies, as the epitome of a rigged system. But Ugland House is not a patch on Delaware (population 917,092), which is home to 945,000 companies, many of which are dodgy shells. Miami is a massive offshore banking centre, offering depositors from emerging markets the sort of protection from prying eyes that their home countries can no longer get away with. The City of London, which pioneered offshore currency trading in the 1950s, still specializes in helping nonresidents get around the rules. British shell companies and limited-liability partnerships regularly crop up in criminal cases. London is no better than the Cayman Islands when it comes to controls against money laundering. Other European Union countries are global hubs for a different sort of tax avoidance: companies divert profits to brass-plate subsidiaries in low-tax Luxembourg, Ireland and the Netherlands." *The Missing $ 20 Trillion: How to Stop Companies and People Dodging Tax, in Delaware as well as Grand Cayman, The Economist*, Feb. 13, 2013, http://www.economist.com/news/leaders/21571873-how-stop-companies-and-people-dodging-tax-delaware-well-grand-cayman-missing-20.

lege should not inhibit the implementation of the FATF Recommendations (Recommendation No. 9); financial institutions should be required to undertake customer due diligence and to verify the identity of the beneficial owner, and should be prohibited from keeping anonymous accounts or those bearing fictitious names (Recommendation No. 10); financial institutions should also be required to maintain records for at least 5 years (Recommendation No. 11) and closely monitor politically exposed persons (PEPs), that is, persons who have greater facility to launder money, such as politicians (in high posts) and their relatives (Recommendation No. 12); financial institutions should monitor wire transfers, ensure that detailed information is obtained on the sender, as well as beneficiary, and prohibit transactions by certain people pursuant to UN Security Council resolutions, such as resolution 1,267 of 1999 and resolution 1,373 of 2001, for the prevention and suppression of terrorism and its financing (Recommendation No. 16); designated nonfinancial businesses and professions (DNFBPs), such as casinos, real estate offices, dealers in precious metals or stones, and even attorneys, notaries, and accountants, must report suspicious operations, and those who report suspicious activity must be protected from civil and criminal liabilities (Recommendation No. 22, in combination with Nos. 18 through 21); countries should take measures to ensure transparency and obtain reliable and timely information on the beneficial ownership and control of legal persons (Recommendation No. 24), including information on trusts—settlors, trustees, and beneficiaries (Recommendation No. 25).

Thus, the true beneficial owners of companies must be collected and made more readily available to investigators in cases of suspected wrongdoing. The costs to openness will be outweighed by the benefits of shining light on the shady corners of finance.

There is much intelligence work to be done, more than that involved in simply controlling one's borders. Intelligence forces need to work together, because if they are kept apart, each may, in isolation, feel that someone else is responsible for the problem.

Along these lines, Terry Goddard informs us that in the Arizona Financial Crimes Task Force searches for financial anomalies, disproportionate events unconnected with economic reality. "They immediately saw that Arizona was a huge net importer of wired funds. At the top-10 Arizona wired-funds locations, over $ 100 were coming in for every dollar wired out. Wire transfers into Arizona from other states, in amounts over $ 500, totaled more than $ 500 million per year. Since there was no apparent business reason for this imbalance, the investigators took a closer look."[3]

In Brazil, foreign-exchange legislation spells out a number of issues that are often unheard of, even in the USA. According to the rules of Brazil's Financial Intelligence Unit, Council for Financial Activities Control—COAF, the item "Cash Transfers" covers remittances and only applies to the mail and Brazilian postal money

[3] Terry Goddard, *How to Fix a Broken Border: Follow the Money*, Tucson Sentinel, Oct. 13, 2012, http://www.tucsonsentinel.com/opinion/report/101312_goddard_border/how-fix-broken-border-follow-money/.

orders, both domestic and international, since everything coming from abroad and involving currency exchange operations comes under Central Bank supervision.

There are operators and currency brokers in the USA who often make use of the gray market to allow transfers of money belonging to uninformed foreigners residing in Brazil. To stock its operations, Brazilian currency (reals), usually in cash and acquired from illegal conduct in Brazil, is deposited by the currency exchange into the accounts of beneficiaries of wire transfers coming from abroad, while the dollars or euros received from the senders, who are easy prey, are diverted to redeem and deposit money as part of this bartering in funds. This is the so-called wire operation.[4]

This is why irregular deposits show up in the accounts of recipients of wire transfers with no identification of the depositor, cash transfers to a beneficiary account from a company not authorized by the Central Bank to operate in the currency exchange market, or cash transfers to beneficiary accounts from individuals, because only financial institutions are eligible to receive Brazilian Central Bank authorization to operate in currency exchange markets.

Note that due diligence is not required when the sending company has its home office in the USA (where that is the source of the dollars).

In response to technological advances and consumer demand to build a global gambling market on the Internet, members of the US Congress have repeatedly attempted to pass legislation prohibiting Internet gambling. On October 13, 2006, President George W. Bush signed into law the Unlawful Internet Gambling Enforcement Act of 2006 (UIGEA)[5], which prohibits the acceptance of payment of wagers by financial institutions. So, the UIGEA bans Internet gambling by forcing financial institutions to prevent financial payments of wagers from bank accounts and other financial instruments.

The UIGEA serves as a monetary control scheme in the USA which intends to prevent the use of credit cards and other bank instruments for Internet gambling. The *raison d'être* for this legislative effort is purportedly to prevent money laundering.

Credit cards[6] that allow access to an account through magnetic media and a password are quite familiar. However, there are prepayment vehicles that may be

---

[4] Something similar to what occurs in the Black Market Peso Exchange, which has long served international drug traffic, also applies to Brazil, with the establishment of the black market in reals. According to Resolution No. 13 of the Council for Financial Activities Control (COAF), factoring companies should report to COAF. The COAF considered important to make it possible to identify the owners and directors of factoring companies, perform due diligence on customers, and check whether internal controls are in place. The aforementioned Resolution was revoked because the Central Bank did not accredit factoring companies as financial institutions, and so they never applied for licenses or registration.

[5] 31 U.S.C. §§ 5361–67 (Supp. 2007).

[6] According to Susan Ormand, "Although online gambling sites accept debit cards, checks, and wire transfers, the predominant form of payment is credit card. Some large banks and credit card companies, including Visa and Mastercard, have refused to transfer money to offshore betting accounts because of the potential for fraud, but bookmakers circumvented these efforts by encouraging bettors to use debit cards or payment services like PayPal, Net-Teller, or E-CashWorld. Conscious of these loopholes and the risk involved, many credit card companies have abandoned

transferred and recharged, and quite possibly put to use by money launderers. Very low identification requirements on the part of financial institutions encourage such criminal practices. They do not themselves store value, but do provide access to an account. It is difficult to draw a distinction between traditional credit cards and the network of prepaid access cards. Stored value cards ought to be clearly classified for the elucidation of government agencies and to facilitate identification of suspicious cards.[7]

An alternative mechanism exists in the form of electronic money[8], and gamblers began using it to place bets or play casino-style games on the Internet. "This alternative method of payment is completely anonymous, untraceable, and more secure than a credit card...perhaps the most powerful and untraceable money-laundering tool ever imagined by criminals."[9]

Electronic money is a digital representation of money that can be placed on a computer hard drive, smart card, or other device with memory, including cellular phones and other electronic communication devices. It can be as anonymous as cash, so there has to be some method or means of breaking that anonymity and tracking the flow of money. A consumer purchases e-money with a form of conventional money or credit. The e-money can then be stored on a smart card or memory-based electronic device until the consumer is ready to spend it. With computer-based e-money, a government or private business issues an electronic coin or note that represents a claim against the issuer and can be redeemed in exchange for traditional money. Once the coin or note has been issued, it can be used online over wires or wireless technology. Unlike credit cards, this coin or note can be used without the assistance of a bank or other traditional financial institution.[10]

For instance, Bitcoins[11] are digital money issued and transmitted over the Bitcoin network, that is, a network among friends which allows payment by one party to an-

---

this extremely profitable market entirely." Susan Ormand, *Pending U.S. Legislation to Prohibit Offshore Internet Gambling May Proliferate Money Laundering*, 10 Law & Bus. Rev. Am. 447 (2004), pp. 451–52.

[7] Such cards are usually not listed among monetary instruments, nor are they otherwise subject to customs declarations, although they often exceed the limit established for a Suspicious Activity Report. Authorities appear oblivious to the need to monitor stored value cards. These innocent cards or instruments can be worth millions, yet authorities are not concerned with them. The cards closely resemble traditional credit cards, but provide access not to credit at a financial institution, but to a sum of money stored on the card, on a chip, or simply in an account accessible using the card (which sometimes even dispenses with the chip).

[8] It is a digital representation of money that can be placed on a computer hard drive, smart card, or other device with memory, including cellular phones and other electronic communication devices.

[9] See, Mark D. Schopper, *Internet Gambling, Electronic Cash & Money Laundering: The Unintended Consequences of a Monetary Control Scheme*, 5 Chap. L. Rev. 303 (2002), p. 304.

[10] See, *id.*, pp. 304, 314.

[11] The bitcoin network began on January 3, 2009, with the issuance of the first Bitcoins by Satoshi Nakamoto. Owners transfer bitcoins by sending them to another bitcoin address using a client program or website for the purpose. The transfer is accomplished by digital signature and connected to the public encryption key of the next owner. Bitcoin records all data necessary to make the transaction valid in the block chain.

other. These transactions are connected to the Internet, and therefore communicate with other equipment, picking up or transferring signals from different regions. The signals are recorded in a public history listing (called a block chain) once validated by the system. This currency runs into two difficulties. The coins are traceable using their unique serial numbers (much handier than the series of numbers printed on bills), and it is not easy to spot someone using the same coin twice because, consisting as they do of numbers, one could make as many copies as one wished. A number of systems have developed cryptographic techniques to prevent such duplication, complicating the transfer of funds to the point of making it difficult to ensure security.[12]

The use of fake names or nicknames, to avoid identifying its users, does not seem impossible. Still, one can easily imagine someone structuring an anonymous payment structure with many branches in order to hide transactions. Electronic money has a potential use for money laundering, offshore banking, and tax havens.[13]

The mere fact that it is electronically traceable does not preclude it from being successfully used for illegal ends. As more of these types of services become available online, the transactions become more complex and there is greater opportunity for apparently unrelated off-the-grid value exchanges. "Electronic money payment schemes, which currently consist of smart cards and computer-based e-money, allow for the storage and redemption of financial value. Simply put, electronic money is a money replacement based on encryption technologies which disguises the electronic information so that only the intended recipient can access its meaning."[14]

NGOs, as well as trusts, associations, and foundations tend to be as diverse as a country's population. People are increasingly getting involved in some kind of social organizations or charity efforts. Donations to these social entities have been large. Just as the work of beneficent entities has been important, scandals have stained some of their images and opened the door to an increase in judicial actions against their directors, increasing skepticism in proportion to news coverage of events.

Because their work is philanthropic, and they are generally motivated by altruism and compassion, charities have been immune to legal proceedings. They may be made answerable on account of internal management issues or even external problems (everything from labor suits to fraud, and even money laundering by reason of insolvency, negligence, or poor practices). Entire boards of directors might be held liable for some failure of accounting or diversion of funds.

Recent disclosures have tarnished the images of certain entities and brought the glare of publicity onto the conduct of some of their managers. A backlash of skepticism has brought about a proportional reaction affecting the volume of donations and volunteer work.

---

[12] *Bitcoin Security*, http://www.snell-pym.org.uk/archives/2011/05/12/bitcoin-security/.

[13] Jon Matonis, *Thoughts on Bitcoin Laundering*, May 13, 2011, http://themonetaryfuture. blogspot.com/2011/05/thoughts-on-bitcoin-laundering.html.

[14] See, Mark D. Schopper, *Internet Gambling, Electronic Cash & Money Laundering: The Unintended Consequences of a Monetary Control Scheme*, 5 Chap. L. Rev. 303 (2002), p. 314.

Because philanthropic work is normally motivated by feelings of generosity and empathy, charitable organizations often imagine themselves immune to legal proceedings. Liability could surface based on some poorly handled internal activity, or some other cause occurring outside the organization. This is why the role of the manager is so important.

By the simple fact that they operate with personal and institutional donations, charity organizations (temples, churches, mosques, NGOs, educational associations, etc.) often believe that they are not required to reveal the source of their funds, nor to be examined for the large financial transactions they conduct.

Managers and employees of NGOs must be answerable for their management and for the protection of the goods and services that benefit us all.

A primary responsibility is to ensure proper accounting for social programs and funding received from their supporters (public or otherwise). This means they must strictly comply with the law and ethical standards, be committed to the mission of the NGO they represent, protect the rights of their members and, indirectly, of those assisted, and prepare annual reports for their country's federal revenue service and regulatory authorities having jurisdiction—reports that should be available to all interested parties.

They should, therefore, have technical information at their fingertips to enable them to monitor and record all assets and amounts received, spent or entrusted to their care.

The website of the National Association of State Charity Officials (NASCO[15]) contains important information on recording and reporting required of NGOs. NASCO members are employees of US government agencies charged with regulating NGOs and their funds.

Marion R. Fremont-Smith, who teaches Public Policy at the John F. Kennedy School of Government, produced an important comparison for Harvard University on the bookkeeping requirements for such organizations. She showed, for instance, that most US states (for example, New York, California, Arkansas, Missouri, and New Jersey) require that they have at least three directors.[16]

FATF Recommendation No. 08, in the spirit of clearly delimiting the rights and responsibilities of directors and employees of NGOs, encourages countries to establish good policy whereby information on their activities, size, and other important characteristics such as transparency, integrity, openness, and best practices can be had in real time for purposes of supervision and monitoring.

NGOs, lacking proper controls, are recognized today as channels for money laundering for organized crime. Money laundering is usually carried out using a layered structure to give the appearance of legality. One such method would be to establish companies through which the company manages business for its clients,

---

[15] www.nasconet.com. Accessed June 2, 2012.

[16] In *The Search for Greater Accountability of Nonprofit Organizations. Summary Charts: State Nonprofit Corporation Act Requirements and Audit Requirements for Charitable Organizations* (document obtained on May 16, 2012, from Patrícia Lobaccaro, president and CEO of the Brazil-Foundation Foundation).

the beneficiary being one or more holding companies, or a series of companies in several tax havens, to create a separation between the aforesaid holding companies and their ultimate beneficiary. Moreover, the discovery of the real beneficiary would require considerable cooperation on the part of authorities in those tax havens. Some means would have to be established to require the trust to provide its beneficiaries' names whenever requested by the authorities.

It is not even easy to establish who is in charge of the trust or the beneficial owner, for there is no obligation that that name be revealed. Hence, being its legal beneficiary is an enviable business—which may explain the rather timid recovery of illegal assets.

NGOs, though they may be temples, churches or mosques, or educational or sport institutions, must reveal sources of their funds and have their financial transactions closely scrutinized.

Considerable customer due diligence is required for their managers, as well as knowing their investors. The donor's name or job/company title is no longer enough. The source of funding must also be disclosed and supported by documentation. In the case of a donor company, a copy of the bylaws is required (in order to check the list of directors) from the civil or deed registry having jurisdiction. The true directors of the NGO must be known, with a photo ID, along with the scope of their authority, all of that backed by documentation to properly support the information provided.

In conclusion, the methods of blossoming untraceable payments, notably virtual ones (cash, e-money) and the use of offshore accounts and NGOs provide unprecedented potential for criminal elements seeking to launder ill-gotten gains in the sport and gambling industries. If, usually under a bank secrecy law, banks and other financial institutions are required to record and report financial transactions to the federal government, it must be mandatory to eliminate or regulate complicated methods of disguising the origin of money so that it appears to be derived from legal activity.

All the efforts to ban or prevent money laundering, or other crimes, can be threatened by technologies (like the Internet and electronic cash) or the use of armored information kept by offshore companies and NGOs.

Bibliography

ANISTIA poderá coibir o 'esquema Daslu'. Colecionadores afirmam que sonegação continuará sendo praxe no meio enquanto carga tributária não diminuir. *Folha de São Paulo*, Jan. 1, 2012, Ilustrada supplement, p. E5.

ARTISTS and Art Galleries. *Internal Revenue Service. Department of the Treasury. Market Segment Specialization Program.* www.artchain.com/resources/art_audit_guide.pdf. Accessed May 23, 2012.

BLACKHURST, Chris. *The Off-shore World of Tax Havens*, 5 Int'l Fin. L. Rev. 16, 18 (1986).

BRIEFING explores the factors that have influenced increases in remitting. *Inter-American Dialogue*, Mar. 20, 2012.

BRINA CORRÊA LIMA, Osmar. Lei das Sociedades por Ações: permanência, mutações e mudança. Revista da Faculdade de Direito da Universidade Federal de Minas Gerais, Belo Horizonte, no.40, pp. 219–32, July/Dec 2001.

CAMARANTE, Andre. Carro da Polícia Civil é atingido por tiros na zona leste de SP, June 22, 2012. www.1.folha.uol.com.br/cotidiano/1109151-carro-da-policia-civil-e-atingido-por-tiros-na-zona-leste-de-sp.shtml.

CAMARGO VIDIGAL, Geraldo de, and SILVA MARTINS, Ives Gandra (Coordenação). Colaboradores: Eros Roberto Grau e outros. Comentários à lei das sociedades por ações (Lei 6404/76). São Paulo: Resenha Universitária: em co-edição do Instituto dos Advogados de São Paulo, 1978.

CANTIDIANO, Luiz Leonardo. Breves comentários sobre a reforma da Lei das Sociedades por Ações. Revista Forense, vol. 360.

CAPPELLI, Silvia. Responsabilidade penal da pessoa jurídica em matéria ambiental: uma necessária reflexão sobre o disposto no art. 225, § 3º, da Constituição Federal. Revista de Direito Ambiental, São Paulo, vol. 1, pp. 100–106, Jan/Mar 1996.

CARTIER, Marie-Elizabeth et al. Entreprise et responsabilité pénale. Paris: LGDJ, 1994.

CHAUVIN, Francis. La responsabilité des communes. Paris: Dalloz, 1996. (Série Connaissance du droit).

CHRISTIE'S. http://christies.com/features/guides/buying/pay-ship.aspx. Accessed May 6, 2012.

CHRISTIE'S CATALOGUE. New York, Old Master Paintings. Wednesday 6 June 2012. London: Christie, Manson & Woods Ltd., 2012.

CINTRA, Marcos Antônio. O Acordo de Basiléia e os bancos públicos. Folha de São. Paulo, São Paulo, Jan. 4, 2006.

COELHO, Fábio Ulhoa. Curso de direito comercial. 4th ed., vol. 1. São Paulo: Saraiva, 2000.

_____. O empresário e os direitos do consumidor. São Paulo: Saraiva, 1994.

_____ et al. Comentários ao Código de Proteção ao Consumidor. São Paulo: Saraiva, 1991.

CORREIA, Eduardo. A responsabilidade jurídico-penal da empresa e dos seus órgãos (ou uma reflexão sobre a alteridade nas pessoas colectivas à luz do direito penal). In: CORREIA, Eduardo et al. Direito penal econômico e europeu: textos doutrinários. Vol. 1. Coimbra: Coimbra Ed. 1998.

_____, and PEDRAZZI, Cesare. Direito Penal Societário. 2nd ed. São Paulo: Malheiros, 1996.

_____. Direito penal das sociedades anônimas. In: ANTUNES, Eduardo Muylaert (Coord.). Direito penal dos negócios: crimes do colarinho branco. São Paulo: Associação dos Advogados de São Paulo, 1990.

CUOMO, Andrew M. Internal Controls and Financial Accountability for Not-for-Profit Boards. Charities Bureau. http://www.oag.state.ny/bureaus/charities/about.html. Accessed June 20, 2012.

_____. Right From the Start—Responsibilities of Directors and Officers of Not-for-Profit Corporations. http://www.oag.state.ny.us/bureau/charities/guide_advice.html. Accessed June 20, 2012.

DENICOLA, Robert. C. Access Controls, Rights Protection, and Circumvention: Interpreting the Digital Millennium. 31 Colum. J.L. & Arts 209, Winter 2008.

DE SANCTIS, Fausto Martin. Money Laundering Through Art. A Criminal Judicial Perspective. Heidelberg: Springer International Publishing Switzerland, 2013.

_____. Responsabilidade Penal das Corporações e Criminalidade Moderna. São Paulo: Saraiva, 2009.

_____. Combate à Lavagem de Dinheiro. Teoria e Prática. Millennium: Campinas, 2008.

_____. Punibilidade no Sistema Financeiro Nacional. Campinas: Millennium, 2003.

_____. Responsabilidade penal das pessoas jurídicas. São Paulo: Saraiva, 1999.

EIZIRIK, Nelson. Instituições financeiras e mercado de capitais—jurisprudência. Vol. 1–2. Rio de Janeiro: Renovar, 1996.

FERRARA, Francesco. Le persone giuridiche. Con note di Francesco Ferrara Junior. 2nd ed. Torino: UTET, 1958. (Trattato di diritto civile italiano, dir. Filippo Vassali, 2).

FOLEY, Rita Elizabeth. *Bulk Cash Smuggling. United States Attorney's Bulletin*. Washington, DC, vol. 55, no. 5, Sept 2007.

FREMONT-SMITH, Marion R. *The Search for Greater Accountability of Nonprofit Organizations. Summary Charts: State Nonprofit Corporation Act Requirements and Audit Requirements for Charitable Organizations* (document obtained on 05/16/2012 from Patrícia Lobaccaro, president and CEO of BrazilFoundation).

GALERIE FURSTENBERG v. Coffaro, 697 F. Supp. 1282 (S.D.N.Y. 1988).

GIORGI, Giorgio. *La dottrina delle persone giuridiche o corpi morali.* 3rd ed., vol. 1. Firenze: Fratelli Cammelli, 1913.

GODDARD, Terry. *How to Fix a Broken Border: Follow the Money*, Tucson Sentinel, Oct. 13, 2012, http://www.tucsonsentinel.com/opinion/report/101312_goddard_border/how-fix-broken-border-follow-money/.

GOVERNMENT plans 'umbrella law' to tighten scrutiny and regulation of religious trusts and NGOs. *Economic Times* (India). Bennett, Coleman & Co. Ltd. May 3, 2011. www.westlaw.com. Accessed 22 June 2012.

HIDALGO, Rudolph et al. *Entreprise et responsabilité pénale.* Paris: LGDJ, 1994.

LEONARDO Jr., Maurício Fernandes. Transferências Financeiras de Pessoas Físicas entre Estados Unidos e o Brasil. Delivered to the author in June 2012. Article undergoing publication.

LINN, Courtney J. One-Hour Money Laundering: Prosecuting Unlicensed Money Transmitting Businesses Using Section 1960. *United States Attorney's Bulletin*. Vol. 55, No. 5. Washington, DC, Sept 2007.

LOBO, Jorge. Curso de Direito Comercial. Rio de Janeiro: Forense, 2002.

MARTINS, Fran. Curso de Direito Comercial. 28th ed. Rio de Janeiro: Forense, 2002.

MATONIS, Jon. *Thoughts on Bitcoin Laundering*, May 13, 2011, http://themonetaryfuture.blogspot.com/2011/05/thoughts-on-bitcoin-laundering.html. Accessed March 8, 2012.

MEXICO proposes to limit cash purchases of certain goods to 100,000 pesos. 2010 Fintrac Report.

ORMAND, Susan. *Pending U.S. Legislation to Prohibit Offshore Internet Gambling May Proliferate Money Laundering*, 10 Law & Bus. Rev. Am. 447 (2004).

PALAK, Shah. Trusts, NGOs under ambit of money-laundering law. *Business Standard*, Mumbai. 2009 WLNR 23270783, Nov. 18, 2009.

PALLANTE, Maria. Symposium: Digital Archives: Navigating the Legal Shoals Orphan Works, Extended Collective Licensing and Other Current Issues. 34 Colum. J.L. & Arts 23 (2010).

PARIENTE, Maggy et al. Les groupes de sociétés et la responsabilité pénale des personnes morales. In: *La responsabilité pénale des personnes morales.* Paris: Dalloz, 1993.

PEDRAZZI, Cesare. O Direito Penal das Sociedades e o Direito Penal comum. *Revista Brasileira de Criminologia e Direito Penal.* Vol. 9. Rio de Janeiro: Instituto de Criminologia do Estado da Guanabara, 1965.

RAJA D, John Samuel. Ten means to put an end to black money issue. Economic Times (India). Bennett, Coleman & Co., Ltd., *The Financial Times Limited*. Nov. 18, 2011.

ROCHA, Fernando A. N. Galvão da. Responsabilidade penal da pessoa jurídica. *Revista da Associação Paulista do Ministério Público*, São Paulo, no. 18, May 1998.

ROCHA, Manuel Antônio Lopes. A responsabilidade penal das pessoas colectivas—novas perspectivas. In: CORREIA, Eduardo et al. *Direito penal econômico e europeu:* textos doutrinários. Vol. 1. Coimbra: Coimbra Ed., 1998.

ROTHENBURG, Walter Claudius. *A pessoa jurídica criminosa.* Curitiba: Ed. Juruá, 1997.

SCHOPPER, Mark D. Internet Gambling, Electronic Cash & Money Laundering: The Unintended Consequences of a Monetary Control Scheme, 5 Chap. L. Rev. 303 (2002).

SEXER, Ives. Les conditions de la responsabilité pénale des personnes morales. Droit et patrimoine, [S.l.], pp. 38–46, Jan. 1996.

SHAH, Palak. *Trusts, NGOs under ambit of money-laundering law*. Business Standard, Nov. 19, 2009, 2009 WLNR 23270783. www.westlaw.com. Accessed June 23, 2012.

SHECAIRA, Sérgio Salomão. A responsabilidade das pessoas jurídicas e os delitos ambientais. *Boletim do IBCCrim*, São Paulo, ed. esp. no. 65, April 1998.

SIKARWAR, Deepshikha. Religious trusts, non-profit organisations to face greater scrutiny. *Economic Times* (India). 2011 WLNR 19624448, Sept. 27, 2011.

SIRVINSKAS, Luís Paulo. Questões polêmicas atinentes à responsabilidade penal da pessoa jurídica nos crimes ambientais. *Revista da Associação Paulista do Ministério Público*, São Paulo, no. 17, April 1998.

SNELL-PYM, Alaric. Bitcoin Security. http://www.snell-pym.org.uk/archives/2011/05/12/bitcoin-security/. Accessed August 3, 2012.

THE Missing $ 20 Trillion: How to Stop Companies and People Dodging Tax, in Delaware as well as Grand Cayman, The Economist, Feb. 13, 2013, http://www.economist.com/news/leaders/21571873-how-stop-companies-and-people-dodging-tax-delaware-well-grand-cayman-missing-20.

TIEDEMANN, Klaus. Responsabilidad penal de personas jurídicas y empresas en derecho comparado. *Revista Brasileira de Ciências Criminais*, São Paulo, no. 11, pp. 21–35, Jul/Sept 1995.

_____. *Poder económico y delito* (Introducción al derecho penal económico y de la empresa). Barcelona: Ariel, 1985.

_____. *Delitos contra el orden económico*: la reforma penal. Madrid: Instituto Alemão, 1982.

VAROTO, Renato Luiz Mello. Da responsabilidade penal dos sócios. *Repertório IOB de Jurisprudência*, São Paulo, no. 2, pp. 30–31, Jan. 1996.

Chapter 7
Combating Money Laundering Through Sport and Gambling: International Legal Cooperation

Depriving financial criminals, such as white-collar criminals, drug and weapons traffickers, racketeers, and members of other criminal syndicates of their ill-gotten proceeds and instrumentalities of their trade not only achieves important law-enforcement objectives but also provides an effective means of recovering funds for victim restitution. The confiscation of such resources by the state also strengthens public safety in general, by allowing for federal investment in state and federal police forces, and in education to prevent serious crime.[1]

Money laundering is usually carried out using a layered structure to give the appearance of legality. One such method is to establish companies that manage business for clients and have one or more holding companies in several tax havens as beneficiaries. This creates a separation between the aforesaid companies and their ultimate beneficiary. Moreover, the discovery of the real beneficiary would require considerable cooperation on the part of authorities in those tax havens. Some means would have to be established to require the trust to provide the beneficiaries' names whenever requested by the authorities.

In the case of money laundering and financing of terrorism, the victim is the government and society as a whole.

International cooperation continues to be an important requirement, considering the increasing globalization of money laundering and terrorist financing threats. The Financial Action Task Force (FATF) has enhanced the scope and application of international cooperation between competent authorities and financial groups through its recommendations. The revised Recommendations will mean more effective exchanges of information for investigative, supervisory, and prosecutorial purposes. This will also assist countries in tracing, freezing, confiscating, and repatriating criminal assets.

While facilitating commerce and communication, the Internet also facilitates the ability of criminals to elude the laws of any, and every, nation. The Internet provides

---

[1] Despite the vast quantities of goods routinely seized and confiscated because of international cooperation, the subject has not held the attention of scholars—unlike other issues such as the length of prison terms, the growing number of inmates, and the quality of penitentiaries and prisons. See, Catherine McCaw, in Asset Forfeiture as a Form of Punishment: A Case for Integrating Asset Forfeiture into Criminal Sentencing. 38 American Journal of Criminal Law 181, 2011, p. 183.

F. M. De Sanctis, *Football, Gambling, and Money Laundering,*
DOI 10.1007/978-3-319-05609-8_7, © Springer International Publishing Switzerland 2014

individuals worldwide with the ability to communicate and exchange information across national boundaries and continents. The Internet has united the globe with access to information, available anywhere, at any time. It has also connected criminals and people with criminal purposes.[2]

Because of its very nature as a network-based technology and the fact that it is not aligned with geography, the Internet requires multinational oversight to provide effective enforcement. It means users have access to sites and information without regard for, or hindrance from, the territorial origin of that data. Consequently, so long as some jurisdictions are willing to allow Internet gambling sites, these virtual casinos are not subject to effective regulation by any one nation, or even a group of nations, because a site permitted in just one jurisdiction is accessible from all others.

As stated in Chap. 3, the attempt of FIFA to obtain vital information through the Transfer Matching System (TMS)[3] is effective but not enough. It is a vital tool for obtaining information about the international transfer of players, previously restricted to only business stakeholders. But efforts by FIFA, which sometimes focus on purely commercial and private interests, should not replace the work of authorities. Certain obligations should be established, like requiring clubs, federations, and confederations—and those who provide advisory, auditing, bookkeeping, and consulting in this area—to communicate suspicious transactions to the Financial Intelligence Units. Clubs, according to the FATF, are deliberately being used to launder money, and thus more must be done. FIFA data are not public and difficult to obtain, and therefore authorities will be forced to request International Legal Cooperation to access the data, because FIFA is headquartered abroad.

In addition to the importance of getting information, it is essential for authorities to track down the assets obtained from criminal activities in sport and gambling.

Confiscation has become a priority strategy in the fight against organized crime. It is especially effective because criminals consider the forfeiture of the equivalent of the proceeds from their criminal activities as upsetting, if not more so, than imprisonment. However, as criminal activity has become transnational, criminal investments have increased incredibly outside of national boundaries. A vast network has emerged to make use of the proceeds of crime and has taken root in loopholes or legal hurdles found in crime-fighting efforts.

---

[2] See MILLS, Jon. In *Internet Casinos: A Sure Bet for Money Laundering*. 19 Dick. J. Int'l L. 77 (2000–2001), p. 83.

[3] Through the Transfer Matching System, over 30 types of information are recorded online, such as player history, clubs involved, payments, values, contracts, etc. It records the information received from both the buyer club and the seller club. For instance, one can check whether the contracted amounts were allocated directly to the parties involved. It is very important that the information is concentrated electronically, but FIFA should not be the only recipient of such data, even though autonomy is guaranteed to the sport by the Brazilian Constitution (such autonomy is limited to the organization and running of the sport). It is essential to create certain obligations, like communications of suspicious transactions to the Financial Intelligence Units, to be imposed on clubs, federations, and confederations, and on those who provide advisory, auditing, bookkeeping, and consulting in this area. There are records of money laundering with the involvement of clubs in the negotiation of international money transfers in various countries. Clubs, according to the Financial Action Task Force (FATF), are deliberately being used to launder money.

Criminals often know how to work the system. This is what justifies shoring up asset forfeiture even for assets that may have been transferred to some third party, who nevertheless ought to have perceived that these were the proceeds of unlawful conduct and for assets transferred to a good-faith third party, precisely to avoid confiscation.[4] The burden of proof is on the third party to show legitimate possession when the government alleges that the item was transferred to him in order to dodge confiscation, or in the belief that he would deliberately act without due caution. This is how repatriation has been accomplished.

All of the money or assets transferred to the state will certainly result in special crime-fighting and prevention programs, as well as combined efforts on the part of the states and the federal government and its agencies, culminating in proper enforcement methods to curtail crime. The ideal is to create a safe place for freedom, security, and justice. This is no small feat, for it necessarily implies a joint effort on the part of the police, prosecutors, and the courts to obtain property illegally acquired. Forfeiture, confiscation, and repatriation require cooperation. Learning from decisions made in the searching out and seizing of the proceeds of crime will result in improvement of the work and training of public agents, and in the adoption of new measures to fight crime.

When assets are confiscated, the message that is sent to organized criminals is that crime no longer pays and is not in the best interests of its practitioners. It drives home the message that some types of behavior are really prohibited, and that to insist on engaging in them is to pay the price.

Confiscation differs from pecuniary penalties, because for pecuniary penalties the criminal violation determines the punishment. But, when assets are forfeited, the amount varies in accordance with the proceeds derived from the criminal behavior. The greater the proceeds from crime, the larger the confiscation is. In other words, ill-gotten goods are taken in proportion to criminal behavior.

Pecuniary fines and confiscation are measures independent of one another, yet incorrectly distinguished by many jurists. Because they are calculated differently, they cannot add those together for purposes of establishing the penalty imposed, for that would reduce the penalty.

The system as a whole works best when confiscation is kept as one of the solutions among several outcomes of sentencing.

In addition to criminal cases, there is also the possibility of administrative or civil confiscation. FATF Recommendation No. 4 makes it clear that no prior criminal conviction is required for the forfeiture of assets.

This occurs frequently in the USA (called actions *in rem,* that is, proceeding against the thing itself), when property or its possession is related to criminal activity.[5] As these claims are framed, ownership or possession is considered illegal, in which case the interested party may intervene to prevent loss to the government.

---

[4] See: Confiscation and Asset Recovery: Better Tools to Fight Crime. States New Service, Brussels, March 12, 2012. www.lexis.com. Accessed May 26, 2012.

[5] Cf. United States v. Bajakajian, 524 U.S. 321, 330–31 (1998) and United States v. One-Sixth Share, 326 F.Wd 36, 40 (1st Cir. 2003).

In such cases, it is incumbent upon the government to prove beyond a reasonable doubt that the asset is actually subject to forfeiture, inasmuch as the thing and its possessor are both somehow connected to crime.[6]

In harmony with this are Article 12(a) of the UN Convention against Transnational Organized Crime and Articles 54 and 55 of the UN Convention against Corruption, which provide for confiscation of the proceeds of, and instrumentalities used in, criminal activities, even in the event of the death of the accused or expiration of the statue of limitations.

FATF Recommendation No. 30 establishes the possibility of conducting freezing and seizure operations, even when commission of the antecedent crime may have occurred in another jurisdiction (country). It also recommends the implementation of multidisciplinary groups or task forces specialized in financial or asset investigations. Thus, international cooperation is essential to vitalize all necessary efforts to assist states in their mission to fight crime.

According to Catherine McCaw, civil and criminal confiscations are not mutually exclusive. Were the government to file both actions, each on its own grounds, and lose one, this in no way precludes continuing to seek confiscation on the remaining action.[7]

No burdens should be placed on polices and state's attorneys' efforts to temporarily freeze or solicit the freezing of assets likely to disappear if nothing is done, and all such measures are, to be sure, subject to consideration by the courts, nor is any such interference warranted when the assets are located abroad.

Governments ought, therefore, to allow the freezing, seizure, confiscation, and repatriation of assets in order to facilitate the fight against organized crime, itself a global business, in such a manner as to force the criminals to change their ways.

According to Fletcher Baldwin, Jr., in examining this ongoing fight against international money laundering, there are three dimensions involved: consistent policies between national and international efforts, an efficient legal and institutional apparatus, and close cooperation between the public and private sectors.[8]

For an effective and cohesive universal policy against money laundering, participation by and commitment of all states to international cooperation is imperative.

The burden of proof is on the third party to show legitimate possession when the government alleges that the item was transferred to him in order to dodge confiscation, or in the belief that he would deliberately act without due caution. This is how repatriation has been accomplished.

Hence, confiscation and repatriation are two stages of legal proceedings in which the assets of criminals are forfeited on behalf of the victims, communities, or gov-

---

[6] 18 U.S.C. § 983 (c)(1)(3).

[7] Catherine McCaw (in Asset Forfeiture as a Form of Punishment: A Case for Integrating Asset Forfeiture into Criminal Sentencing. 38 American Journal of Criminal Law 181, 2011, p. 195).

[8] See Fletcher Baldwin, Art Theft Perfecting The Art of Money Laundering. (Jan. 2009 for the 7th Annual Hawaii International Conference on Arts & Humanities). An unpublished work, sent to the US Library of Congress on April 20, 2012, by University of Florida College of Law Professor Emeritus Levin, by request of the author, pp. 47–48.

ernments. Central to this procedure is the decision that a given asset was acquired as the proceeds of unlawful conduct, and may therefore be confiscated.

The first stage in repatriation is the tracking and identification of goods. This normally involves coordinated efforts on the part of prosecutors and government agencies (revenue authorities, police, and private collaborators). These efforts also require substantial expertise and skill in dealing with the financial transactions that are sometimes involved. International legal cooperation has been essential to shedding light on the activities of organized groups. It has enabled the blocking of goods and repatriation of assets, which invariably rely on companies or institutions headquartered in tax havens or elsewhere abroad.

One approach calls for the application of reciprocity, according to which governments may cooperate with one another in the absence of some previous treaty or international agreement, acting through mutual commitments undertaken in dealing with a specific case.

Provisions for mutual reciprocity are found in numerous national and international laws and treaties, including: the UN Convention against the Illegal Traffic in Narcotic Drugs and Psychotropic Substances (Vienna, 1988, Articles 6 and 7), Law No. 11343 dated August 23, 2006 (the Brazilian drug law, Article 65), the Extradition Treaty to which MERCOSUL member states are party[9], the UN Convention against Transnational Organized Crime (Palermo, 2000, Articles 16–19), the UN Convention against Corruption (Mérida, 2003, Articles 44 and 46), the Council of Europe Convention on Laundering, Search, Seizure and Confiscation of the Proceeds from Crime (Strasbourg, 1990, Articles 7–35), the Council of Europe Convention on Laundering, Search, Seizure and Confiscation of the Proceeds from Crime and on the Financing of Terrorism (Warsaw, 2005, Articles 15–45), Brazil's Money-Laundering Law (No. 9613, of March 3, 1998, Article 8, as amended by Law 12683 of July 9, 2012), and the Model Regulation promulgated by the Inter-American Drug Abuse Control Commission (CICAD/OAS, Article 20).

The FATF clearly emphasizes in its Recommendations the need to reinforce international cooperation through general exchange of information relating to suspicious transactions. There is the understanding that the various standards relating to the element of intent in criminal conduct should not affect the ability or the will of countries to cooperate on judicial matters. The Recommendations establish the possibility of freezing and seizure even where the antecedent crime is committed in some other jurisdiction (country), as well as the implementation of specialized multidisciplinary teams or task forces (Recommendation No. 30); international legal cooperation, pursuant to the UN Conventions of Vienna (international traffic, 1988), Palermo (transnational organized crime, 2000), and Mérida (corruption, 2003), by withdrawal of obstacles (Recommendation No. 36); direct mutual legal assistance to enable a quick, constructive, and effective solution (Recommendation No. 37);

---

[9] They specifically address International Legal Cooperation and one conclusion arrived at in Rio de Janeiro on 12/10/1998 was promulgated in Brazil through Legislative Decree No. 605 of 09/11/2003, which became effective internationally on 01/01/2004 as a result of the Treaty of Asunción, creating the South American common market, MERCOSUL (signed 03/26/1991).

freezing and confiscation even where there is no prior conviction (Recommendation No. 38); extradition (Recommendation No. 39); and competent authorities that can rapidly and effectively provide international cooperation in relation to money laundering, antecedent crimes, and terrorism financing (Recommendation No. 40).

International cooperation, however, requires more than just legal cooperation. It also requires so-called administrative cooperation, not contingent upon indictments. In the latter case, all communication occurs through intelligence channels. Information exchanged through direct cooperation between Financial Intelligence Units, the Attorney General Offices, and police authorities in many countries.

As Patrícia Núñez Weber explains:

> International administrative cooperation is in the strictest sense not tied to any criminal demands or occurrences, but aimed at technological improvement, exchange of information, creation and maintenance of databases, and the sharing of strategies among the agencies involved. Yet the term is also used to designate cooperation among administrative authorities quite apart from court orders.[10]

This brings us to the possibility of direct exchanges of information, through the aforementioned intelligence channels. Most of the information would originate out of legal cooperation, under the aegis of the Judicial Branch, most notably in cases that require measures such as seizure and lifting of bank or tax record secrecy, that is, whenever a court order is needed.

The Internet, because of its very nature as a network-based technology, requires multinational oversight to provide effective enforcement. It is not aligned with geography, and users have access to sites and information without regard for, or hindrance from, the territorial origin of that data. Consequently, so long as some jurisdictions are willing to allow Internet gambling sites, these virtual casinos are not subject to effective regulation by any one nation, or even a group of nations, because a site permitted in just one jurisdiction is accessible from all others.

It is important to mention that the bureaucracy and procedural delays inherent in processing letters rogatory lead to an increase in another mode of international cooperation, direct assistance. This allows us to get around the sending and procedural delays of letters rogatory, for it allows direct transmission. Direct assistance has emerged as an effective alternative in the fight against international crime. Through this form of cooperation, authorities other than the judiciary avail themselves of international requests, and the procedures are much simpler than those involving traditional letters rogatory and even dispense with the prima facie evaluation in Brazil.

This brings us back to the observations of Patrícia Núñez Weber, who explains:

> Direct assistance is cooperation offered by national authorities and likely to satisfy the foreign request, in the performance of their legal duties as though it were a national procedure, when in fact it arises from a request by a foreign State channeled through Brazil's central authority. (…)
> Currently, the most widespread understanding is that direct assistance presupposes the existence of a treaty or agreement with the requesting State, or a promise of reciprocity. Our

---

[10] *Apud* Carla Veríssimo De Carli (Org.). *Lavagem de dinheiro: prevenção e controle penal.* Porto Alegre: Verbo Jurídico, 2011, p. 589.

feeling is that that restriction arises from the relatively recent arrival of the institution on the international scene, compared to letters rogatory.[11]

In direct assistance, a request is received by the central authority and then forwarded to the judiciary. The judge may then examine the facts presented by the foreign nations on their merits, much as in domestic proceedings to which procedural rules would apply. This is something that does not happen with letters rogatory.

Requests for direct assistance are generally couched in terms of international treaties or agreements. One approach is through the application of reciprocity, according to which governments may cooperate with one another in the absence of some previous treaty or international agreement, acting through mutual commitments undertaken in dealing with a specific case.

For example, José Antônio Dias Toffoli and Virgínia Charpinel Junger Cestari explain:

> Requests for direct assistance are, as a rule, couched in terms of bilateral treaties or agreements (the so-called Mutual Legal Assistance Treaties or MLATs). Absent any express understanding between the two States assistance can still be provided based on the requester's assurance of reciprocity. This allows cooperation in many different tax, labor or pensions-related areas. Still, the treaties most frequently encountered in an international setting have to do with criminal and civil subject matter.[12]

The institution of "Central Authority" came about to speed up and facilitate cooperation between countries. As the name itself suggests, the primary role of the central authority is to function as a centralizing agency, the focus of all cooperation—requests and investigations alike—whether coming from abroad or transmitted from within the country. All letters rogatory and requests for legal assistance, whatever their purpose, should be handled through the intermediation of the Central Authority.

In the USA and Brazil, as in most countries, the Central Authority lies in the Executive Branch (respectively, the Department of Justice and Ministério da Justiça), given that it typically represents the state in international relations. The Central Authority is an idea espoused in the Hague Conventions and international conventions on public international law, arising out of the need to have an agency in each country to regulate the administrative procedures for International Legal Cooperation.

Its creation was imperative, given the increase in volume and complexity of mechanisms for international cooperation. It imparts uniformity of performance, standardizes all procedures, and provides the necessary specialization for handling such matters, avoiding duplication and waste in the requests.

There are countless advantages to the institution of a single Central Authority: specialization, speed, efficiency, publicity, and affordability of proceedings. It is

---

[11] *Apud* Carla Veríssimo De Carli. *Lavagem de dinheiro: prevenção e controle penal.* Porto Alegre: Verbo Jurídico, 2011, pp. 593 and 602.

[12] In *Mecanismos de Cooperação Jurídica Internacional no Brasil.* Manual de Cooperação Jurídica Internacional e Recuperação de Ativos—Matéria Civil. Asset Recovery and the International Legal Cooperation Council Department, National Secretariat of Justice, Ministry of Justice. 2nd ed. Brasília: 2009, p. 27.

argued that placing the Central Authority in the Executive Branch will also ensure neutrality, transparency, and due process, inasmuch as Executive Branch agencies are subject to oversight by the Office of the Public Prosecutor, their acts subject to review by the Judicial Branch.[13]

Cooperation may occur directly between any of the competent authorities. However, the institution of Central Authority brings these authorities closer together to eliminate obstacles in the way of rapid realization of shared national interests. Indeed, there is no point in demanding the establishment of a Central Authority, unless it is committed to achieve efficiency, simplification, and necessary speed of information and action. Yet, it is important to mention that the Central Authority is by no means a sine qua non for making international cooperation feasible.

Freezing and seizure operations require significant efforts and hard work. It is not enough to simply attach a court order. At times, one has to turn over convincing documents to establish a link between assets or a bank account and illegal activity. It helps if the goods in question are the proceeds from criminal activity abroad or at least flow (by action or omission) from corrupt practices. Hence, legal assistance has allowed the freezing and repatriation of assets, but often requires an affidavit—an internally consistent sworn statement—to enable such measures as freezing assets or bank accounts.[14]

One interesting case in the USA had an impact on US legislation regarding asset forfeiture. The case involved Brazil filing for freezing of assets belonging to a Brazilian defendant, and for keeping those assets in the USA. The issue was whether, based on 28 U.S.C. § 2467(d)(3), titled Enforcement of Foreign Judgment[15], foreign

---

[13] The great challenge at this point is to further popularize the benefits of adopting a single Central Authority in each country for all International Legal Cooperation issues, and to broaden the horizon of this institution. With the assistance of the policies of Brazil's Justice Ministry, through the National Anti-Money Laundering Qualification and Training Plan (PNLD), the idea is being spread that, even in the absence of an agreement, it is possible to have requests for active or passive cooperation routed through the Central Authority.

[14] Brazilian Congressman Eduardo Valverde submitted a bill (No. 1982, dated September 16, 2003) to regulate International Legal Assistance on Criminal Matters irrespective of the transmittal of letters rogatory. The bill, which is still pending in the Brazilian Congress, would provide for temporary administrative freezing of proceeds of crime undergoing laundering. It also would provide a Council on International Legal Assistance empowered to formulate directives and serving as a permanent clearinghouse for information among the various government agencies it represents (the Federal Courts, Office of the Federal Prosecutor, Ministry of External Relations, Office of the Attorney General, Brazil's Federal Revenue Secretariat, the Central Bank, the Council for Financial Activities Control (COAF), the Federal Police Department, and the Office of the Comptroller-General), all offering guidance to Brazilian authorities needing to secure international cooperation.

[15] 28 U.S.C. § 2467(d)(3) states as follows:
Entry and Enforcement of Judgment—(1) In general: The district court shall enter such orders as may be necessary to enforce the judgment on behalf of the foreign nation unless the court finds that:

(A) The judgment was rendered under a system that provides tribunals or procedures incompatible with the requirements of due process of law;
(B) The foreign court lacked personal jurisdiction over the defendant;
(C) The foreign court lacked jurisdiction over the subject matter;

assets may be frozen only after a foreign court has definitively ruled in favor of forfeiture, or if it may be done before any final decision on confiscation has been rendered. The United States Court of Appeals for the District of Columbia Circuit, on review of two lower-court decisions from March and April 2009, decided that a final decision by Brazil regarding confiscation was required, according to its interpretation of 28 U.S.C. § 2467(d)(3).[16] Following this decision, the US Department of Justice requested and obtained from the Congress a resolution of the problem because, if upheld and followed, the decision would have compromised international cooperation efforts with other countries.

By the principle of specialization, applicable to relations among states and, therefore, to international legal cooperation efforts, no information or documents obtained through legal assistance may be used with regard to crimes for which international cooperation is excluded on account of jurisdiction being regarded as an attribute of the state. Switzerland provides such an example, with regard to exchange quota violations.[17]

Insistence on dual criminality, that is, that the alleged behavior constitutes a crime in both of the jurisdictions involved in the request for legal assistance on criminal matters is a common requirement in cooperation cases—with the proviso that the two criminal categories need not be exactly the same, but only similar. Indeed, the UN Convention on Corruption signed at Mérida makes it clear that "in matters of international cooperation, whenever dual criminality is considered a requirement, it shall be deemed fulfilled irrespective of whether the laws of the requested State Party place the offence within the same category of offence or denominate the offence by the same terminology as the requesting State Party, if the conduct underlying the offence for which assistance is sought is a criminal offence under the laws of both State Parties" (Article 43).

The rule cannot, however, be interpreted as an absolute. The FATF recommends international legal cooperation, pursuant to the UN Conventions of Vienna (international traffic, 1988), Palermo (transnational organized crime, 2000), and Mérida (corruption, 2003), by withdrawal of obstacles (Recommendation No. 36) and direct mutual assistance toward a quick, constructive, and effective solution (Recommendation No. 37). Thus, not even the FATF gives the rule the proper crucial weight with regard to money laundering. Failure to honor the principle of *dual criminality,* which is no novelty in international public law, may prove devastating in future

---

(D) The foreign nation did not take steps, in accordance with the principles of due process, to give notice of the proceedings to a person with an interest in the property of the proceedings in sufficient time to enable him or her to defend; or

(E) The judgment was obtained by fraud.

[16] United States v. Opportunity Fund and Tiger Eye Investments, Ltd., 613 F.3d 1122 (D.C. Cir. 2010).

[17] Pursuant to ENCLA 2005 Target No. 40, the Justice Ministry's Asset Recovery and International Legal Cooperation Council Department agreed to share information on the need to keep within limitations on the use of documents obtained through International Legal Cooperation, and reaffirmed the principle of specialization at the international level.

requests for cooperation, in specific cases, and completely bar new freezes or obtaining of evidence, among other diplomatic difficulties.

With specific regard to the sharing of goods between requesting and requested states, as observed above, the UN Convention signed at Mérida (corruption) made no provisions, however, there was an understanding that there ought to be full restitution of assets to the injured-party-state in view of the legal assets affected (Articles 51–59).

Finally, with regard to seizure or freezing of assets, there are no obstacles to international application once the universal rule of reciprocity is in effect, and it is still possible to share assets confiscated or seized, and consequently repatriate them, once the decision awarding forfeiture to the government becomes final. It is extremely difficult to obtain repatriation of assets based only on an appealable decision, even if reciprocity is invoked.

The authorities of requested states usually wish to be informed regarding: (1) evidence that all owners, agents, curators, or others involved with the articles to be seized are aware of issuance of the order that they be seized, and of its content; (2) evidence to show that the seizure order was signed prior to the legal sale or transfer of the articles abroad; (3) proof of direct association between the article and fraud which would demonstrate that its acquisition does indeed flow from criminal behavior; (4) unavailability of the assets (e.g., works of art) precisely because such a procedure is public knowledge and one might therefore infer that the interested parties, curators, art dealers, and such had knowledge of the illegal events involving large holdings; and (5) listing of all legal events (seizure orders, forfeitures, decisions that have become final) relating to the accused, including all corresponding dates.

At the first Judicial Roundtable Meeting between federal judges from Brazil, the USA, Colombia, and Mexico[18], in Washington, DC, October 27–31, 2011, initial conclusions were drawn that the more the evidence, the easier the seizure/sequestering, and that where drug trafficking is concerned, it is easier to secure the granting of the request because the laws are not as difficult to understand. The purpose of this unprecedented encounter was to establish a framework of judicial decisions to render International Legal Cooperation feasible as quickly as possible.

In conclusion, international cooperation is a prerequisite to prevent money laundering through gambling and sport, especially with the use of Internet. Because of the implications of online dealings for funding criminal endeavors, every country must assist in opposing money laundering in other jurisdictions through international cooperation.

---

[18] Two judges per country, along with US federal prosecutors, attended the meeting organized by the US Department of Justice.

Bibliography

BALDWIN, Fletcher. *Art Theft Perfecting the Art of Money Laundering.* (Jan. 2009 for the 7th Annual Hawaii International Conference on Arts & Humanities). An unpublished work, sent to the U.S. Library of Congress on April 20, 2012 by University of Florida College of Law Professor Emeritus Levin, by request of the author, pp. 47–8.

CESTRI, Virgínia Charpinel Junger, and TOFFOLI, José Antonio Dias. Mecanismos de Cooperação Jurídica Internacional no Brasil. *Manual de Cooperação Jurídica Internacional e Recuperação de Ativos*—Matéria Civil. Asset Recovery and the International Legal Cooperation Council Department, National Secretariat of Justice, Ministry of Justice. 2nd ed. Brasília: 2009.

CONFISCATION and Asset Recovery: Better Tools to Fight Crime. *States New Service*, Brussels, March 12, 2012. www.lexis.com. Accessed May 26, 2012.

DE CARLI, Carla Veríssimo (Org.). *Lavagem de dinheiro: prevenção e controle penal.* Porto Alegre: Verbo Jurídico, 2011, p. 589.

LOULA, Maria Rosa Guimarães. *Auxílio direto: novo instrumento de cooperação jurídica internacional civil.* Belo Horizonte: Fórum, 2010.

MADOFF, Diana B. Henriques. Apologizing is Given 150 Years, *New York Times*, June, 30, 2009, A1.

MAZZUOLI, Valério de Oliveira. *Curso de Direito Internacional Público.* 2nd ed. São Paulo: Ed. Revista dos Tribunais, 2007.

MCCAW, Catherine E. Asset Forfeiture as a Form of Punishment: A Case for Integrating Asset Forfeiture into Criminal Sentencing. 38 American Journal of Criminal Law 181, 2011.

MERCED, Michael J. De la. Prosecutors Try to Claim Madoffs's Properties, *New York Times,* March 17, 2009, B6.

MILLS, Jon. *Internet Casinos: A Sure Bet for Money Laundering.* 19 Dick. J. Int'l L. 77 (2000–2001).

PINHEIRO, Luís de Lima. *Direito internacional privado.* Vol. 1. Coimbra: Almedina, 2002.

REZEK, Francisco. *Direito internacional público:* curso elementar. 10th ed. São Paulo: Saraiva, 2005.

THE UNITED STATES DEPARTMENT OF JUSTICE, Asset Forfeiture Program, Annual Financial Statements, FY 2011 Report no. 12–12. http://www.justice.gov/jmd/afp/01programaudit/index.htm. Accessed May 25, 2012.

UNITED STATES v. Bajakajian, 524 U.S. 321, 330–31 (1998).

UNITED STATES v. One-Sixth Share, 326 F.Wd 36 (1st Cir. 2003).

UNITED STATES v. Opportunity Fund and Tiger Eye Investments, Ltd., 613 F.3d 1122 (D.C. Cir. 2010).

Chapter 8
Conclusions

The fight against crime requires much more than its mere discovery. There are five stages for this process: (1) prevention, (2) repression, (3) trial and judgment, (4) recovery, and (5) reintegration. Schools, families, churches, and nongovernmental organizations (NGOs) are responsible for the first stage, followed by the police (the second), the justice system (the third), and by the justice system and other government sectors (fourth and fifth stages). The judiciary plays an important role, especially in the third stage. If the best solution to the case is not found—that is something visible only from a perfect global overview of the problem—the judiciary could come to be perceived as nearly useless, a bit of theater.

Federal agencies, prosecutors, and judges must fight crime in a coherent and systematic way. This is the expected existential minimum of the institutions, for which truth is a value intrinsic to the judicial function. It cannot be applied solely to the parts (prosecution or defense), subject as they are to the imperfections of humanity: laziness, selfishness, dereliction, and disdain.

On a par with seeking a proper judicial decision, prevention is carried out by the society to keep serious events from coming to the courtroom, or delaying too long in arriving there due to the dereliction that accompanies a preventive system that does not work.

Casino games, lotteries, and sports are global activities prone to criminal activity and money laundering because of the large sum of money channeled into them. If vulnerabilities and anonymity persist in these areas, the risk of exploitation by organized criminals will continue to grow.

Discussions about the need to reform pertinent legislation should be unanimous as to the direction things ought to be taken. Trends toward a regulatory environment of extreme liberty do not necessarily lead to proper protection of vital assets, and even less do they reflect procedural guarantees in criminal law. Even the leading proponent of the theory of guarantee, Luigi Ferrajoli, defended excessive regulatory freedom.[1] To the established doctrinaire, criminal guarantees, meaning the Con-

---

[1] Ferrajoli's "garantismo," with some foothold in constitutional maxims, reinforcing principles of *nulla poena sine juditio* and *nulla poena sine processu*, among others, fell short of nullifying the quest for real truth through excessive resort to appeals or curtailing the court's exercise

F. M. De Sanctis, *Football, Gambling, and Money Laundering,*
DOI 10.1007/978-3-319-05609-8_8, © Springer International Publishing Switzerland 2014

stitutional Rule of Law, are those set of duties and rational rules imposed upon all powers in the exercise of protection over the rights of all, which is the only proper response to dog-eat-dog regimes, a response that divides guarantees into primary and secondary. "Primary guarantees are the limits and regulatory ties—that is, prohibitions and obligations both formal and substantive—imposed in the protection of rights, on the exercise of any power. Secondary guarantees are the various forms of reparation—the capacity to nullify wrongful acts and all liability for illegal acts—which themselves flow from violations of primary guarantees."[2]

It is important to establish a clear and systematic set of rules for the prevention of money laundering on the cases already reported, besides the methods verified on them.

The judicial branch of government, in interpreting the Constitution, is no mere executor of rules conjured up by the will of common legislators, but rather acts as the guardian of fundamental rights.

Criminal law, like any and all other laws, in seeking a solution for conflict situations, with the necessary weighing of the values at stake, leaves all concerned satisfied with the result of a claim almost invariably adverse to one of the parties. Its mission is therefore delicate and boils down to a redefinition of the prosecution, requalifying it oftentimes by adoption of consequences most damaging to strictly personal property.

The regulation of financial law, especially financial criminal law and here we are generally referring to the idea of the war on money laundering (especially in the ill-disciplined world of gambling and sport), is justified by the simple idea that market rules alone cannot provide all of the aspirations emerging within the context of the course of business practices—oftentimes through dangerous, ethical gray areas.

There is a need to fill the loopholes which ordinary criminal law has largely been insufficient to properly suppress given the increase in financial crime arising out of the exponential increase in international crime.

If financial crime has often elicited lackadaisical social reaction, because of a perception that this is something engaged in by the people of social or professional prestige, we have seen that in many court cases, the opposite emerges to the extent that money laundering (especially using gambling and sport) has arisen out of ordinary serious crimes, such as drug trafficking.

According to Graham Johnson, "money laundering is the key to a modern crime empire and is carried out clinically, professionally, and sustainably."[3]

---

of jurisdiction over a substantive case. Nor has it come down to point-by-point analysis of facts placed before it. It comes down to determining necessity, fault-based liability, adequate proof, cross-examination, and the right to a hearing. Then again, one does not go off in search of truth at any price, especially at the cost of historic advances which have resulted in the recognition of universal rights, consisting at first of individual principles, progressing through collective, to transindividual—the right of the whole.

[2] Cf. Ferrajoli, Luigi. *El garantismo y la filosofía del derecho*. Bogotá: Universidade Externado de Colombia, p. 132.

[3] *See* Graham Johnson. *Football and Gangsters: How Organized Crime Controls the Beautiful Game*. Great Britain: Cox and Wyman Ltd., 2007, p. 199.

The absence of overt violence could perhaps explain some moral neutrality. I believe, however, that society is slowly becoming aware of the deleterious effects of financial crime, for it in turn also fosters ordinary crime. We must move in the direction of imprisoning these financial criminals, who are themselves increasingly dangerous and bold. Hence, on account of the cost–benefit calculation of the proceeds of criminal behavior versus possible outcomes (penalties) imposed by the legal system, they ought not to be dealt with leniently or with less than the strictness that is called for.

It is necessary to keep in mind that criminals must ultimately be reintegrated into society. Therefore, the application of pecuniary sentences or curtailment of rights must be considered and measures must be adopted to close legal and institutional loopholes through which crime perpetuates itself.

It is a tricky analysis, easily swung toward the trivialization of human rights, and the serious danger of the rights being seen so irrelevant that any conclusion from the balance of their meaning is self-disqualifying. Criminal law for enemies, as advanced by Günther Jakobs,[4] or criminal law for enemies as a "third speed," as advanced by Silva Sánchez,[5] would serve to preserve not only the law but also people's confidence in the criminal justice system, so that the purpose of the penalty is to reaffirm the rule of law.

There is a regulatory and functional concept of culpability (reproachability), without ontological foundation, and is based on general positive prevention.[6] Hazards facing the foundations of a punitive system are acquiescence based on potentialities (preemption of punishability based on future events rather than actual occurrences), the imposition of disproportionate harsh penalties, and nonobjectivity of procedural guarantees. This is criminal law divorced from context, only to meet the expectations of general prevention: a dangerous instrumentality, violative of human dignity.[7]

The idea here is neither to defend this theory nor to nullify criminal law through the practical impossibility of its enforcement. Even worse would be to perpetuate the general impression that this branch of law penalizes poor criminals while catering to the wealthy. That would be frustrating, unjust, evil, and in its consequences, a danger to democracy and to belief in democracy.

Hence, a clear and systematic ordering of existing rules on money laundering, especially laundering made possible through gambling or sport, so as to block off

---

[4] Cf. JAKOBS, Günther, and MELIÁ, Manuel Cancio. *Direito penal do inimigo. Noções e críticas*. Transl. André Luís Callegari and Nereu José Giacomolli. Porto Alegre: Livraria do Advogado, 2005. pp. 66–69.

[5] *Id.*

[6] *Autonomy (the subject's capacity to abide by law) is held to be a functioning capacity where its outcome is functional, and absent that there must be another way of resolving the conflict.* (ROXIN, Claus. *Derecho penal—Parte general—Fundamentos. La estructura de la teoria del delito*. Madrid: Civitas, 2006. t. 1, p. 806).

[7] We would have violations of human dignity, and the subject would come to be used by the state to further its preventive convenience. (ROXIN, Claus. Reflexões sobre a construção sistemática do direito penal. *RBCCrim* 82/35).

any possibility of criminal behavior, must be defended. Here, then it would make sense to put every effort into effective crime fighting not only restricted to setting up regulations for the sector but also aimed at the improvement of payment methods and a clearer notion of the kind of work done by NGOs.

This book had the scope to review, strengthen, and establish measures that enable reflection on some laws and regulations in gambling and sport in order to improve financial transparency and curtail organized crime.

Legislators, regulators, scholars, and authoritative bodies must reflect on how money launderers operate. When officials do not effectively control gaming and sport, the varied opportunities for money laundering increase in number. Exchange of currencies, using cash, money transfers, etc., becomes vulnerable to criminal activity.

According to John Warren Kindt and Stephen W. Joy, "an increased risk of white-collar crime… [was] predicated upon the ease with which gambling facilities laundered money. This factor, coupled with the anonymity of the Internet and lax supervision by government officials in certain countries hosting on line casinos, resulted in a strong likelihood of criminal behavior eventually occurring."[8]

Awareness, regulation, and effective controls ultimately protect such activities, the economy, and the health of the concerned businesses. They allow the continuation of an adequate and fair competition, innovation, and industry growth.

Some countries, such as Brazil, opted to ban casinos (Decree-Law no. 9.215/1946), unlike others such as the USA, which restricted the ban to Internet Web sites. To those that adopt absolute prohibition, the rule applies to all who are established in the country, including local Web sites.

The people involved in games and sports, including the employees and the players themselves, must act in accordance with the standards of honesty, integrity, and fairness, so that it is possible to ensure proper performance of the business and serve the public interest.

The regulation aims to provide effectiveness to investigations in order to be able to promote, indirectly, sustainable growth of the industry through procedures and controls that enable regular collection of resources and taxes.

Governments still have weak and vulnerable anti-money laundering policies, providing remarkable mechanisms for the commission of crimes, especially financial ones.

Failures in recognizing suspicious activities and the lack of obligation to report them to authorities are now unjustifiable due to the importance of combating money laundering.

Any rule authorizing or regulating games or sports must include provisions for tax payments with specific purposes (health, safety, culture, and sport). It would be relevant to institute a monthly monitoring fee and permit financial intelligence units (FIUs) to check financial transactions online and in real time. These FIUs

---

[8] JOY, Stephen; WARREN KINDT, John. Internet Gambling and the Destabilization of National and International Economies: Time for a Comprehensive Ban on Gambling Over the World Wide Web. 80 Denv. U. L. Rev. 111 2002–2003, p. 119.

should be well structured and staffed with people with professional and technical qualifications, who are able to effectively control a supposed beneficiary's entities. The creation of the National Register of Compulsive Gamblers would prohibit their persistence of attending such environments.

Cases mentioned in this book and the methods used in the crime demonstrate that it is more than necessary to establish strict controls over such activities, which should be really transparent.

Regarding sport, "stating that sport participation is not detrimental for society is not equivalent to saying that sport participation is surely beneficial for society."[9]

The analysis of football in its darkest aspect demonstrates that dribbles, "hats," incredible goals, packed stadiums, parties, and children and adults with their eyes transfixed before their idols, have been shaken by football's connection with organized crime, illegitimate, or illegal commissions for transfers of players, and unlawful profit from its finance arm. The cross-border flow of money via payments through questionable means (cash/e-money, wire transfers, use of offshore accounts, and NGOs) that has characterized the negotiations within the sector has not been properly disclosed. A greater financial control is required in order to eliminate loopholes that criminals take advantage of, a phenomenon that multiplies the instances of delinquency. The field of criminal financial offenses is fertile and threatens the grass field.[10]

The Brazilian Special Account Court, called Tribunal de Contas da União, has the authority to help Congress review public expenses of the federal government. Its last report about the World Cup in 2014, delivered in April 2013, revealed that due to its work, it was possible to save around US$ 600 million.[11]

According to Jorge VI, Alagoas State Sport Secretary and president of the club CSA, the Brazilian Football Confederation would have charged US$ 2.5 million for authorizing the State of Alagoas to host the Confederation Cup in 2013 and the World Cup in 2014. He alleged that all the companies that would participate in

---

[9] *See* Raul Caruso. Relational Goods at Work! Crime and Sport Participation in Italy: Evidence from Panel Data Regional Analysis over the Period 1997–2003. *Contemporary Issues in Sports Economics. Participation and Professional Team Sports.* Cheltenhanm, UK: Edward Edgar Publishing Ltd., 2011, p. 59.

[10] The "Mane Garrincha" stadium in Brasilia, with a capacity for 71,000 spectators (the second largest in Brazil, behind only the Maracanã in Rio de Janeiro), was budgeted at US$ 650 million (US dollars on May 20, 2013) and funded through the negotiation of land negotiated by and belonging to Terracap, property state, whose society is divided between the government of Brasilia (51%) and the Federal government (49%). The directors of the company were all dismissed by Governor Agnelo Queiroz. The final expense is estimated at US$ 1 billion. The Court of the Federal District discovered irregularities of all sorts, such as duplicated costs of leased equipment, oversized transportation vouchers, overpayment for services not performed, and overpricing in some items valued at US$ 36 million. As of now, the Maracanã has cost about US$ 506 million for just a second reform (in A terra do "Nunca fica Pronto". *Revista Época*, edição 781, May 13, 2013. São Paulo: editora Globo, p. 30–35).

[11] See O TCU e a Copa do Mundo de 2014, Brasília, April 2013, in http://portal2.tcu.gov.br/portal/page/portal/TCU/copa2014.

public procurement for future works would be given preference and concluded that the State of Alagoas would not accept misprocurements.[12]

The market is easy to penetrate because of its complex network of investors or shareholders, management by nonprofessionals, diverse law structuring, considerable sums involved, irrational investments, and a constant need for funding clubs. Such conditions are responsible for the compromise of the social role of football and break the illusion of innocence in sports.

Money gradually began to dominate the world of sport due to the advantages that are directly or indirectly obtained. On one hand, the increase in cash flow has allowed large numbers of people to access the world of sport through various investments. On the other hand, it has led to harmful effects. There is a high risk of fraud, corruption, and money laundering.

Crime has demonstrated its adaptability to find ways to launder illicit money and has created a global problem in the effect it has had on sport and football.

The first document from the European Union Committee that recognizes the importance of the sport was published in July 2007 (EU White Paper on Sport). It states that, "sport is confronted with new threats and challenges, as commercial pressures, exploitation of young players, doping, corruption, racism, illegal gambling, violence, money laundering, and other activities detrimental to the sport."[13]

Many factors have led to the use of illegal resources in football given its complex organization and insufficient or negligible transparency. It is easy to access federations and clubs, which historically have had amateur administrations or conveniently shielded administrations.

Social status acts are also an element of attraction and result in questionable sums of money with no apparent or explicable financial return or gain other than the natural prestige of investing through sport. Its popularity can be a fertile tool for criminals to legitimize themselves through appearance alongside famous people or respected authorities.

Football is an obvious candidate for study because of its astounding transformation from a popular sport to a global industry with significant economic impact. Given its social importance, it has been a vehicle for the transmission of cultural and universal values.

Many cases have shown that football covers every kind of illegality, including money laundering, corruption, and even drugs. Drug barons seemingly like to buy football clubs so that they can launder systematically, in large scale, week in and week out.

The lack of transparency regarding the transfer of players and the true owner or manager of football clubs, which are usually in the hands of national associations,

---

[12] Cf. Jorge VI, presidente do CSA-AL denúncia corrupção em obras da Copa do Mundo. *Futebol Interior*, Dec. 13, 2012. In http://www.futebolinterior.com.br/campeonato/alagoano-primeira_divisao-2013/247493+Presidente_do_CSA-AL_denuncia_corrupcao_em_obras_da_Copa_do_Mundo.

[13] Cf. FINANCIAL ACTION TASK FORCE—FATF Money Laundering through the Football Sector Report. In http://www.fatf-gafi.org/topics/methodsandtrends/documents/moneylaunderingthroughthefootballsector.html. Last updated February 1, 2012.

can lead it to domination by a handful of people and cause serious concern about prevention and suppression of money laundering.

To protect children and adolescents, FIFA established that international transfers are possible only after the minor turns 18 years (Article 19 of the Rules on Status and Transfer of Players), but a perfect application of the rule would require proper registration of athletes. Early transfers of players have proven to be a modern form of slavery, and the scope of the FIFA rule seems to inhibit the practice of every kind of criminal offense within football, especially human trafficking, corruption, tax evasion, and fraud.

Direct contact between athletes and clubs should be seen as an essential way to prevent intermediaries or transfer agents from cheating the players. On the other hand, authorities have underestimated the problem or do not address the issue with the attention required to provide an effective financial transparency to prevent tax evasion and capital flight through football. Many techniques include financial transfers with the use of offshore companies located in tax havens, with shelf companies, and with the intermediation of nonprofessional or politically exposed persons.[14]

Clubs are being voluntarily or deliberately used for criminal purposes and authorities are unable to avoid it.

In Brazil, there are no reports of effective analysis of the sector vulnerabilities, despite the fact that traditional bodies like the Brazilian Federal Reserve (Banco Central), the Council for Financial Activities Control (COAF—Conselho de Controle de Atividades Financeiras), Brazilian Securities Commission (CVM—Comissão de Valores Mobiliários), the Brazilian IRS (RFB—Receita Federal do Brazil), federal police, federal prosecutors, and federal courts strive to do their part.

Thus, awareness of the problem contributes to solving it at appropriate levels so that all players can start to understand their liability in fighting illegal activities. There should be, therefore, perfect coordination between the sector and government authorities with ongoing training.

From the analysis of some particular cases, it is evident that money is laundered in sport through unjustified investments into clubs, whether in the form of sponsorship, loans, through the use of front persons, or through overvaluations in the purchase or sale of players (with the appropriation of the difference between the actual value and stated one in the transaction). Such situations require the use of tax havens, the creation of nonexistent companies, and the use of false documentation.

It is important to establish the obligation to report suspicious transactions, especially regarding business transfers of athletes through promotion, brokerage, marketing, or trading. It is also essential to require advice and consultancy in financial matters involving football because oftentimes the criminals are precisely those who would have the obligation of suspicious activity reports.

---

[14] The Normative 748/RFB Revenue Service of Brazil, of June 28, 2007, does not require the full identification of partners and managers, contrary to the legal entities domiciled in the country, but rather requires only the presentation of a copy of the charter, in order to obtain the CNPJ (the identification number for companies in the Brazilian Internal Revenue Service). This may be satisfied with the mere statement issued by a public tax haven (with company name, date of opening, legal, corporate, and address).

Structured financial transactions are often used to take advantage of the lack of transparency in transactions that can multiply monetary flows through intermediaries, especially when they are diverse in number and carried out by border crossing.

Identifying and containing various money laundering schemes that vary in complexity and use a combination of techniques to obscure the money trail is not an easy task.

Also, the use of nonfinancial professionals, such as family members, lawyers, consultants, and accountants as a means of creating structures for the movement of illicit funds has also been observed by the Financial Action Task Force (FATF).

The money stipulated in such image contracts (for exploitation of a player's personal appearance as part of an extensive advertising campaign) is commonly transferred to accounts of companies in tax havens with serious risks of fraud. Advertising and sponsorship contracts can also be used for money laundering. Organized crime could sponsor sport and constitute a bridge to legitimate business. The most common form of payments involves jurisdictions located abroad, always as a way to hide the last destination.

The efforts of FIFA are not enough to prevent unlawful practices. Associations, federations, and confederations must engage and establish proper references or guidelines within football, and provide the necessary support to clubs through professional training in order to allow suspicious transaction reports.

There is not a real concern among authorities to address the issue at the national level and effectively provide the necessary transparency in financial operations.

France, which created the Direction Nationale du Contrôle de Gestion (DNCG), provides an important example of the professional treatment of financial management in clubs. It is comprised of a body of accountants and is concerned about the solvency of clubs and investors and has a supervisory role on the origin of the funds invested in football. Such an experience demonstrates an important way to remove financial control over "clubby" football.

The confidentiality rule cannot justify the refusal of reporting suspicious transactions in the wake of the FATF Recommendation. Indeed, the duty to report suspicious activities by professionals not connected to the financial sector, also in line with the FATF Recommendations, is essential because usually the managers and actors involved in the recruitment of players and their rights are not from the financial sector.

The role of bodies responsible for preventing money laundering is relevant due to the necessary records for the performance of certain investments.

Task forces make their members work in perfect coordination, although members occasionally must respond to their original offices to comply with the rigid formal rules. Thus, it is important to recognize the need for institutions to work together in a coordinated way to enable a quick response to the actions of organized criminals, which historically have relied on the lack of organization of state agencies to intensify their actions.

There is no minor role of the Securities and Exchange Commission (SEC, and in Brazil, the Brazilian Securities Commission or CVM), since its function is to supervise, regulate, and sanction the securities market. The SEC, for instance, must check the accounting profile and verify the possibility of controversial statements of companies open to the public, which are occasionally controlling football teams. These companies have to make their financial accountings following the provisions of law and regulations issued by the SEC.

Investment funds also have an important function in detecting if there are monthly reports to shareholders and if there is expenditure on a regular basis to investors.

In turn, the Internal Revenue Service may have a prominent role in verifying if clubs are complying with the tax revenue rules. It should consider if the immunity granted to sports bodies, incorporated as nonprofit associations, is a subjective immunity associated with the activity of the entity, or an objective one, linked to a specific operation.

The Federal Reserve may exert substantial control over the operations carried on athlete's business passes with foreign associations since it is important to check whether conditions in contracts between clubs were based on regular exchange.

In turn, the financial institutions involved with the money flows to and from abroad should refuse transactions if there is a negative supply information about the owners of the resources.

One of the objectives of FIFA is to prevent practices that could endanger the integrity of sport. That is why it considers global regulation that can charge the incisive action on any people and places necessary.

The FIFA Statutes and its implementing regulation constitute the constitution of the governing body of international football. They determine the basic laws of the football world, including rules about competitions, transfers, doping issues, and a variety of other subjects.

James G. Irving stated that FIFA should act more as the strong leader of this "football family" and that it mishandled the transfer negotiations from the start under Blatter's administration. For him, "one thing is for sure, with the 'football family', nothing is certain and, more significantly, nothing is easy."[15]

The adoption of isolated measures, apart from any global consideration of the problem, will never suffice as an effective strategy in the war against money laundering. Even self-regulated entities, such as FIFA, have understood the seriousness of the problem.

Football confederations, like the Brazilian Football Confederation (CBF), have exercised minimal control over suspicious activities involving football. Their performance is limited—sometimes CBF is in charge of the issuance of the International Transfer Certification for players and is not aware about conditions or the value of transactions. It is true that CBF does not interfere in the transaction between athletes and clubs. For these reasons, the confederations or federations have no way to inform the amounts and contracting parties involved in their loans.

Brazilian law includes certain oversight mechanisms that the sport sector is based on: the autonomy of sports bodies (Article 217 of the Constitution), which may be constituted as business associations; financial and administrative transpar-

---

[15] In *Red Card: The Battle Over European Football's Transfer System*. 56 U. Miami L. Ver. 667, 2001, pp.723–25.

ency; morality in sport management; the social liability of the leaders; the assignment of football as a cultural heritage; the institutional role of federal prosecutors with the obligation of its defense in Federal Courts ("Pelé" Law No. 9,615/1998, with amendments established by Law No. 10,672/2003); the consecration of fan rights to an organized tournament; transparency in regulations and ticket sales; and having safe places for competition and good food and sanitation in stadiums (Statute Defense Fan, Law No. 10,671/2003).

According to Rafael Teixeira Ramos, there are three features of the international sporting autonomy: self-primacy normative, initiatory, and legality.[16] It means that above the Olympic Letter and the Federated International Statutes (e.g., the FIFA Statutes) there is no other, i.e., their grounds are based in coordination with representative bodies of the associative-Olympic-legality sports movement. This primacy would be self-grounding with the legal capacity for self-regulation in order to standardize the understanding of the sport and to defend monopoly of the organization of competitions. Legislation on money laundering should require persons or entities entrusted with the promotion, brokerage, trading, and trading rights of transfer athletes to report suspicious transactions.

The organization and practice of sport should be regulated by a national entity (in Brazil, the Sports National Council), which must have a professional sport composition feature.

Law No. 9,615 of May 24, 1998, the so-called Pelé Law, replaced the "Zico" Law and was later amended by Law No. 9,981 of July 14, 2000, and Law No. 10,671 of May 15, 2003. The "Pelé" Law regulated partnership agreements among clubs, sponsors, and investors. The chapter devoted to "bingo" (gambling) was revoked by Article 2, Law no. 9,981/2000. So this kind of gambling has been prohibited in Brazil since December 31, 2000.[17]

---

[16] Cf. *Direito Desportivo. Tributo a Marcílio* Krieger. Coordenation: Leonardo Schmidt de Bem and Rafael Teixeira Ramos. São Paulo: Quartier Latin, 2009, p. 100–103.

[17] The Pelé Law basically includes the following: sovereignty in the organization of sport, the organizational autonomy, and football as a national identity. It also determines an ethical and transparent financial management, with social liability (Article 2, I and II, and the single paragraph); the obligation of the National Sports System can be established autonomously, requiring quality standards and integrating the Brazilian cultural heritage, considered of high social interest (Article 4, IV §§ 1–3); the obligation of National Sports to supervise the application of the principles set forth, issuing opinions and recommendations, and to propose measures for the application of funds from the Ministry of Sports among others (Article 11, I–V); to obtain public funds, entities administering the sport should perform all acts necessary to allow accurate identification of one's financial situation, provide rescue plans and investments, ensure the independence of their boards of directors, supervisory professional model, and prepare and publish its financial statements as defined in the Law 6404 of December 15, 1976, after being audited by independent auditors (Article 27, § 6). It also provides accountability of managers when they deviate from purposes set, or work for themselves or for third parties. The clubs can be established as a corporation (Article 27, § 9); sporting entities that do not constitute a business company will qualify as a society in common, pursuant to Article 990 of the Civil Code, which obliges the partners to respond jointly and severally by social obligations. Thus, members began to be held jointly liable for the debts of the clubs (Article 27, § 11); prohibition to the same person or entity to control a single team (Article 27A), the incidental nature of the bond with sports the athlete, dissolving it for all legal purposes,

Thus, there is significant regulation of the practice of football in Brazil. Football is seen as part of cultural heritage, with specific constitutional protection.

Making professional football clubs financially transparent would involve publication of their accounts, which would gather all relevant financial information similar to what the DNCG does in France. This would create a positive image of the sector in relation to third parties such as banks, suppliers, and sponsors, besides providing an economic perspective to clubs, either by obtaining finance or sponsorship.

The National Sports Council (CNE) in Brazil is the collective body in charge of deliberation and advice and is responsible to the sport department (Minister of Sport). It fails to have, however, the same authority as the DNCG, because its actions are limited to establishing guidelines, without the power to make more individualized analyses of the economic situation of clubs, federations, and confederations, much less to stop businesses that are able to jeopardize the financial circumstances of them.

The creation of a body similar to the DNCG could, perhaps, constitute an important sign of repudiating malpractice. It should be comprised of not only people from the sport sector but also members of the community, including trade unions, auditors, etc. With a totally independent structure, it should also have members from the federal police and federal prosecutors, in order to have power over the managements of clubs.

Disciplining and standardizing financial statements would facilitate the analysis of the accounts and the auditing of sporting clubs.

There is an international concern about money laundering through football clubs, which can be used for unjustified investments, whether in the form of a sponsorship or a loan, through front companies or individuals. Moreover, there are cases of overvaluation of federal rights,[18] which would cause the appropriation of the

---

valid from March 25, 2001 (Article 28, § 2, and Article 93); the preference for the first renewal of the original contract professional, with a maximum term of 2 years by the athlete's training entity (Article 29, caput); the need to submit independent audits for sports leagues and entities administering sport at the National Sports Council (Article 46A). Law 10,671, of May 15, 2003, amended by Law 12,299, of July 27, 2010, called the Statute of Defense Fan, brought important innovations, notably with regard to administrative and financial transparency of economic activity in sport. It guarantees the right of fans to organized contests with regulations and ticket sales, security in places of the competitions, safe transportation in stadiums, and quality food in stadiums. It makes clear, in Article 4, § 2, that the sports organization is founded on freedom of association and a member of the Brazilian cultural heritage with high social interest. Due to this, the federal courts have jurisdiction according to Article 70 of Complementary Law 75 of May 20, 1993 (Organic Law of the Federal Public Prosecution). This law also established the right to transparency and publicity of sports entities linked to professional football, but its Article 5 was vetoed, so there is only the mandatory disclosure of income (Article 7). By stipulating the guidelines of Sports Justice, the law made it clear that it should be guided by impersonality, speed, publicity, and independence, requiring the course of the process under penalty of nullity (Articles 34–36). Football is contemplated by incentives and tax benefits (Law 11,438 of December 29, 2006, regulated by Decree 6,180 of March 8, 2007).

[18] Federal rights are the right of a club in registering the athlete in the federation as bound to him. Thus, the federative rights arise precisely from the conclusion of the employment contract and cannot be divided. A player may not be registered in the federation for more than one club at a time.

difference between the current value and the one on the contract of transaction. So, it is beneficial to have a legislation that is able to restrain hiring service companies owned by directors or persons connected with the board, athletes, and people related to clubs, federations, and confederations. Although some laws, such as in Brazil, contemplate the need of publication of club accounts, and even punishments for those who carry out fraudulent administration, effective supervision is necessary to enforce such laws.

The risk arising from absent or insufficient supervision is always the possibility to commit crimes, such as Internet crimes, drug trafficking, human trafficking for sexual exploitation of women and children, terrorism, etc. Reducing crime will require the creation of task forces for planning, coordination, and cooperation among all actors responsible for this difficult mission, including educators and NGOs. Policies that take into account only the increase of police forces are not enough.

Sport has created both a justification and an attraction for various criminal activities. The apparent inability of the authorities to deal with it requires a confident public policy to deter wrongdoing.

This book was an attempt to enable the construction of a better awareness of the problem among players, clubs, federations, and confederations. Approaching this difficult issue must be managed in a professional way to improve financial transparency by imposing best practices. Establishing a code of ethics and building cooperation with the private sector are also essential.

The consecration of independence in sport does not diminish the need to recognize, in the football industry, the duty requirements found in any other industry. These include transparency, external audits, financial management, and properly structured and effective regulations.

The possible tension between independence and equal treatment should be attenuated with the awareness of the social, educational, and cultural importance of sports.

This study tried to establish a position on various highly complex and controversial points and hoped to contribute effectively to foster debate or, at least, provide the reader the scope of the problem in an activity that has generated hatreds and passions on one side and gradual channeling of vast sums of money on the other. Its investments not only represent an increase of sports practices but also create a very high risk of fraud, racism, corruption, human trafficking, and money laundering.

Problems related to gambling have attained certain sophistication and involve several operators in different countries and the use of the Internet. Therefore, countries should regulate the gambling market to make it transparent as profiteers have advantage when there is a lack of proper supervision. It is not easy to contain the speculators if they use online services and are established abroad.

The purpose of this book was not to exhaust any situation arising from the underlying crimes committed through gambling or sport, but to seek information on varied topics within the limits of the theme in order to consider the challenges and threats surrounding sport which, unfortunately, has been subject to commercial pressures, young athletes' exploitation, doping, illegal betting, and all sorts of violence.

Sport has required fair play to inhibit common violence, but should demand a financial fair play to curb bad practices, which can compromise and strike down one of the pillars of the culture of many countries like Brazil.

Observed cases of money laundering through gambling and sport were detected by unconnected sectors. This represents an alert to the banking systems, since precedent crime was discovered (tax fraud, capital flight, and corruption, for example) and culminated in arrests, convictions, and confiscation of assets.

Thus, a rethinking of the role of casinos, lottery houses, FIFA, confederations, federations, clubs, players, offshore accounts, and NGOs is necessary to get them to adapt to the real situation, especially to be aware of the suspicious payments that increase the hazards that have hardly been assessed.

Governments should take a tougher stand against money laundering and quickly detect it by closing the loopholes that allow money to flow among organized crime cartels. In view of the foregoing, developing a policy to establish the role of each actor in the system is of utmost importance to facing up to the current problem.

Thus, an outline is sketched for the reader to be able to conduct a critical, real, and practical analysis, in an effort to restore to the society what it deeply and sincerely believes in: its own dignity and cultural heritage.

States, the police, prosecutors, judges, and government agencies are all essential to efforts toward international cooperation and repatriation of goods. Observe that the US courts have been sympathetic to the requests of foreign governments for the return of assets, despite all of the obstacles, criticism, and tension usually surrounding litigation of that sort.

As pointed out by José Paulo Baltazar Júnior "failure to act on the part of a legislator is practically the same, in effect, as counterproductive interference."[19] Thus, the thing that the law is there to protect by positive action by the state must be protected, and the means to do so effectively must be accepted. This is to say that clearly defined government standards are necessary in order to perform existing protective duties (prohibition of insufficiency). This sort of normative construct is not original with this author. It comes from the Federal Constitutional Court of Germany, in a decision on abortion (*BVerfGE*, 39, 1 ff.—*Schwangerschaftsabbruch*) February 25, 1975.[20] The idea is occasionally referred to as a reflection of the prohibition of excess. The theory (prohibition of insufficiency) calls for restriction of fundamental rights of potential perpetrators of aggression, that is, protection through state intervention. Legislators must make an effort to not fall short of minimum protection requirements. The prohibition of excess involves a substantive legislative act (its content) and the suitability of the response involved (the minimum requirement). The discussion of prohibition of insufficiency revolves around the necessity of law (as a precondition) for the protection of basic rights (as opposed to legislative dereliction). Both constitute the right to defense, and are complementary government guarantees of freedom on different levels.

---

[19] BALTAZAR JR., José Paulo. *Crime organizado e proibição de insuficiência*. Porto Alegre: Livraria do Advogado, 2010. pp. 53–54.

[20] *Id.* p. 55.

So the course and bearing to be set are bound to attain satisfactory levels of efficacy only attainable through absolute technical rigor. Lack of technique, casuistry, or losing sight of the principles which guide criminal justice will hinder rigorous application to specific cases.

In a contemporary government, it is inevitable that some new legal understandings will take root, but this has to be accompanied by harmonization, universal application, impersonality, or morality of language drafted, but without losing sight of social reality and established principles.

Hence, the fight against the trafficking of drugs and weapons, corruption, terrorism, and money laundering requires concerted action that is not restricted to the regions affected by the criminal behavior. One task requiring systematic protection from governments is protection of the integrity of international financial systems. Resolution No. 1373 adopted September 28, 2001, by the UN Security Council makes it clear that global stability is threatened by sophisticated criminal organizations, including terrorists, as well as by traditional and modern methods of money laundering.

It is not just the concern of immediate participants in the gambling and sport world, but primarily is of public authorities, including police, government agencies, prosecutors, and even judges. Thus, everyone, including governments, must contribute to profound and broad concerted action. This means cooperation at the national and international levels.

This idea is seldom dealt with or at best timidly advanced in university and institutional circles, whether by professors, students, or industry professionals. Their best efforts are invested in issues supposedly more relevant, although there is already an idea of criminal justice making a greater effort at doing away with serious crime, that is, in the fight against organized crime.

It was not easy to outline a subject so complex and riddled with secrecy. Governments are becoming stronger, more cooperative, and are joining together to fight crime. I do not believe that human rights are being abandoned, but rather, that modern criminality as an offshoot of globalization is now better understood.

The balancing act in this sector, as in all human action, legitimizes a host of public decisions, some already adopted and others being advanced with prudence, responsibility, and public spiritedness. This should be our primary goal. Policies concerned with ideological or economic interests are really no more than powerful individual interests masquerading as the former and must give way before that primary goal.

The emotional involvement in gambling and sports gives us an idea of how transcending the need is to protect these sectors, irrespective of origin and the country to which they belong.

In presenting this work, I have strived to bring out my constant, daily concern over these modern times, times that stand the authorities at defiance and press them to take action—action other than the quest for some sort of abstract happiness which finds its expression in derision or the disgrace of another. Welfare is itself a form of heritage to be protected by all, each contributing with small deeds (never in isolation), most notably those who carry the torch of power for people thirsting for justice.

My purpose was to go beyond a mere introduction to this engrossing subject. The idea was to present additional considerations in an effort to closely study the gambling and sports industries, which will also further the prevention of money laundering in general.

The idea was—to the extent possible—to look outward toward that universe that surrounds our world, in an effort to bring about improvements in systems for law enforcement and crime prevention.

Bibliography

A TERRA do "Nunca fica Pronto." *Revista Época*, edição 781, May 13, 2013. São Paulo: editora Globo, pp. 30–35.

BALTAZAR JR., José Paulo. *Crime organizado e proibição de insuficiência*. Porto Alegre: Livraria do Advogado, 2010.

CARUSO, Raul. Relational Goods at Work! Crime and Sport Participation in Italy: Evidence from Panel Data Regional Analysis over the Period 1997–2003. *Contemporary Issues in Sports Economics. Participation and Professional Team Sports*. Cheltenhanm, UK; Northamgton, MA/ USA: Edward Edgar Publishing Ltd., 2011, pp. 43–62.

FERRAJOLI, Luigi. *El garantismo y la filosofía del derecho*. Bogotá: Universidade Externado de Colombia, p. 132.

FINANCIAL ACTION TASK FORCE—FATF Money Laundering through the Football Sector Report. In http://www.fatf-gafi.org/topics/methodsandtrends/documents/moneylaunderingthroughthefootballsector.html. Last updated February 1, 2012, accessed on May 10, 2013.

GODDARD, Terry. *How to Fix a Broken Border: FOLLOW THE MONEY*. Part III of III. American Immigration Council Publication. Immigration Policy Center, May 2012.

INTERPOL. http://www.interpol.int/en/content/search?searchText=worksofart. Accessed June 21, 2012.

IRVING, James G. *Red Card: The Battle Over European Football's Transfer System*. 56 U. Miami L. Ver. 667, 2001.

JAKOBS, GÜNTHER, AND MELIÁ, MANUEL CANCIO. Direito penal do inimigo. Noções e críticas. Transl. André Luís Callegari and Nereu José Giacomolli. Porto Alegre: Livraria do Advogado, 2005. pp. 66–69.

JOHNSON, GRAHAM. Football and Gangsters. Great Britain/Edinburgh and London: Cox and Wyman Ltd., 2007.

JOY, Stephen; WARREN KINDT, John. *Internet Gambling and the Destabilization of National and International Economies: Time for a Comprehensive Ban on Gambling Over the World Wide Web*. 80 Denv. U. L. Rev. 111 2002–2003.

KOBOR, EMERY. Money Laundering Trends. United States Attorney's Bulletin. Washington, DC, vol. 55, no. 5, Sept 2007.

MARPAKWAR, Prafulla. State forms cells to detect source of terror funds. *Times of India*. Copyright 2011 Bennett, Coleman & Co. Ltd. 12/24/2011. www.westlaw.com.

MATTOS, Rodrigo. Custo oficial da Copa sobre 10% e vai até R$ 28 bilhões. In: http://copadomundo.uol.com.br/noticias/redacao/2013/06/18/custo-oficial-da-copa-sobe-10-e-vai-ate-r-28-bilhoes.htm, accessed June 19, 2013.

MEXICO proposes to limit cash purchases of certain goods to 100,000 pesos. 2010 Fintrac Report.

OLSON, Eric L. *Considering New Strategies for Confronting Organized Crime in Mexico*. Washington, DC: Woodrow Wilson International Center for Scholars. Mexico Institute. March 2012.

O TCU e a Copa do Mundo de 2014, Brasília, April 2013. In HTTP://portal2.tcu.gov.br/portal/
   page/portal/TCU/copa2014, acessed June 15, 2013.
JORGE VI, presidente do CSA-AL denúncia corrupção em obras da Copa do Mundo. *Futebol
   Interior,* published on December 13, 2012, http://www.futebolinterior.com.br/campeonato/
   alagoano-primeira_divisao-2013/247493+Presidente_do_CSA-AL_denuncia_corrupcao_em_
   obras_da_Copa_do_Mundo.
RAJA D, John Samuel. Ten means to put an end to black money issue. Economic Times (India).
   Copyright 2011 Bennett, Coleman & Co., Ltd., *The Financial Times Limited.* November 18,
   2011.
RAMOS, Rafael Teixeira. *Direito Desportivo. Tributo a Marcílio Krieger.* Autoria coletiva. Coordenation: Leonardo Schmidt de Bem and Rafael Teixeira Ramos. São Paulo: Quartier Latin,
   2009.
ROXIN, Claus. *Derecho penal—Parte general—Fundamentos. La estructura de la teoria del
   delito.* Madrid: Civitas, 2006. t. 1, p. 806).
_____. Reflexões sobre a construção sistemática do direito penal. *RBCCrim* 82/35.
THEODORO JÚNIOR, Humberto. *Consumers' Rights (Direitos do Consumidor: a busca de um
   ponto de equilíbrio entre as garantias do Código de Defesa do Consumidor e os princípios
   gerais do direito civil e do direito processual civil).* Rio de Janeiro: Forense, 2009.
TWO TOP Cartels at War in Mexico. *Express.* Washington, DC: a publication of the *The Washington Post,* p. 6, 05/25/2012.

Chapter 9
Proposals to Improve the War Against Money Laundering Through Gambling and Sport/Football

A number of different international and national initiatives are being put forth in the war against money laundering and the financing of terrorism. International treaties, supplemented by recommendations from foreign multilateral organizations, along with recurring discussion meetings, have all sought to improve the global system of enforcement to curb these serious crimes. Now, we turn to effective enforcement in the sectors under study: gambling and sport/football.

The subject is one of the constant concerns, for much is said about the need to improve these sectors, yet in practice, not much has been done toward the prevention of crime. There should be a more significant record of reporting on the part of casinos, lottery houses, confederations, federations, and clubs. The real work ahead is to get all government agencies to put some teeth into the prevention and punishment of money laundering through the recovery of dirty money.

There is no word of important advances in this area, and that may explain the fact that organized crime finds an extraordinary array of techniques to put a legal face on the proceeds of crime using gambling and sport. It is hard not to notice that major drug traffickers, among other criminals, look to the gambling and sport industries as a way of investing the profits from their deals or as currency to use in the illegal drug market.

What is called for is an immediate rereading of all mechanisms of law enforcement and prevention of money laundering as a general proposition, and all of its myriad forms of expression, but notably in this very important area where enforcement is the reaffirmation of cultural and social traits.

This chapter presents proposals to reflect upon. These proposals are not meant to serve as the last word, but rather as the beginning of a new debate over important forms of money laundering. It is hoped that these proposals will prove useful not only to the mentioned sectors but also as a means of better handling prevention and enforcement in the war on money laundering and the financing of terrorism.

F. M. De Sanctis, *Football, Gambling, and Money Laundering,*
DOI 10.1007/978-3-319-05609-8_9, © Springer International Publishing Switzerland 2014

9.1   General Proposals

9.1.1   An International Perspective

9.1.1.1   Financial Action Task Force (FATF)

Explanation The Financial Action Task Force (FATF) has expressed particular concern regarding casinos (Recommendation Nos. 22 and 28), but not about lotteries and sport, despite the significant number of cases discovered so far.

01. Include clearly all forms of gambling and sport in the Suspicious Transaction Reporting Recommendation for nonfinancial companies and professions alongside casinos, real estate brokers, dealers in precious metals or gemstones, attorneys, notaries, and accountants (Recommendation No. 22, together with 18–21).

9.1.1.2   Law Enforcement Agencies, Financial Intelligence Units (FIUs)

Explanation The FATF recommends that all countries identify, evaluate, and understand the hazards they face from money laundering and the financing of terrorism, and that they take coordinated action to mitigate it (*risk-based approach—RBA*, Recommendation No. 1). This would provide for cooperation and national coordination of prevention and enforcement policies, with proper actions, financial intelligence units (FIUs), etc. (Recommendation No. 2). All countries should:

- Establish the requirement of *customer due diligence* (CDD), whether companies or individuals, a ban on anonymous accounts or those bearing fictitious names, and identification requirements for their beneficial owners (Recommendation No. 10).
- Require records to be kept for at least 5 years (Recommendation No. 11).
- Broaden the definition of "politically exposed persons," that is, persons more readily able to launder money, such as politicians and their relatives (in prominent positions) to include both nationals and foreigners, and even international organizations (Recommendation No. 12).
- Require reporting of suspicious operations on the part of designated nonfinancial businesses and professions (DNFBPs) such as casinos, real estate offices, dealers in precious metals or stones, and even attorneys, notaries, and accountants. Internal controls must be established along with protection of whistleblowers from civil or criminal liability (Recommendation No. 22, in combination with Nos. 18–21). Moreover, transparency ought to be required of the beneficiaries of companies, and countries should obtain sufficient information in real time (Recommendation No. 24), including information about trusts, settlors, and trustees or beneficiaries (Recommendation No. 25). Financial information units need to have timely access, direct or indirect, to financial and administrative information in the hands of law enforcement authorities in order to fully perform their

functions. This includes analysis of suspicious activity reports (Recommendation Nos. 26, 27, 29, and 31).

- Properly regulate casinos, with effective supervision and rules to prevent money laundering (Recommendation No. 28). Although the FATF did not publish recommendations similar to those for casinos to cover the gambling and sport industries, this does not mean that the FIUs cannot proceed thus, as indeed the Brazilian intelligence unit saw fit to do, but with no oversight (Council for Financial Activities Control—COAF, Resolution No. 005 of July 2, 1999, for "bingos"[1] and Resolution No. 18, of August 26, 2009, for lotteries).

02. Establish regulations, irrespective of any obligation arising in law (provided there is a basis in the reasons for which they are created), as to the requirement of suspicious activity reports by individuals or companies that sell, import, export, or intermediate a sale—whether on a permanent or temporary basis, in a principal or accessory role, and cumulatively or otherwise, to prevent the laundering of money through gambling and sport, with rules clear enough to include confederations, federations, and clubs. In the absence of a regulatory agency, oversight shall be performed by the FIUs.

03. Require suspicious activity reports on the part of deed registries or by agencies in charge of regulating real estate brokers, most notably when cash payments or attempted cash payments occur, or payments are made through overseas accounts. There are frequent reports of people acquiring real estate and paying for it entirely or in large part in cash, which has caused unprecedented inflation in the real estate market.

9.1.1.3 Tax Havens, Offshore Accounts, and Trusts

Explanation The best way to combat illegal activities is transparency, which permits the collection and sharing of more information. It is important to crack down on the use of nominal shareholders and directors to hide the provenance of money.[2]

---

[1] Brazil, despite being one of the most populous countries in the world (about 200 million), has a small gaming industry. The industry reached its peak in 2006, when there were over 130,000 machines including slot machines in about 1,500 bingos operating across the country. However, in 2007, many bingos were forced to a standstill and machines were confiscated by the authorities when a federal police operation named "Hurricane" revealed rampant corruption in the industry, with suspected involvement of politicians and organized crime in order to keep them open. Any project aiming to reopen bingos would create a serious risk of money laundering if it is accompanied by a large enforcement structure of regulation and oversight mechanisms sufficient to contain organized crime. No justification, including the allocation of revenue to sports, culture, health, safety, opening new jobs, and investment, would legitimize permission unless crime is able to be contained.

[2] According to an article in *The Economist*, "not all these havens are in sunny climes; indeed not all are technically offshore. Mr Obama likes to cite Ugland House, a building in the Cayman Islands that is officially home to 18,000 companies, as the epitome of a rigged system. But Ugland House is not a patch on Delaware (population 917,092), which is home to 945,000 companies, many of which are dodgy shells. Miami is a massive offshore banking centre, offering depositors from emerging markets the sort of protection from prying eyes that their home countries can no longer

The FATF (or *Groupe d'Action Financière sur le blanchiment des capitaux*—GAFI) as stated before, established 40 recommendations that highlight the importance of transparency to avoid money laundering:

- Countries should establish policies to supervise and monitor nonprofit organizations, obtaining real-time information on their activities, size, and other important features such as transparency, integrity, and best practices (Recommendation No. 8).
- Financial institution secrecy laws, or professional privilege, should not inhibit the implementation of the FATF Recommendations (Recommendation No. 9).
- Financial institutions should be required to undertake CDD and verify the identity of the beneficial owner, and should be prohibited from keeping anonymous accounts or those bearing fictitious names (Recommendation No. 10).
- Financial institutions should be required to maintain records for at least 5 years (Recommendation No. 11).
- Financial institutions should closely monitor "politically exposed persons" (PEPs), that is, persons who have greater facility to launder money, such as politicians (in high posts) and their relatives (Recommendation No. 12).
- Financial institutions should monitor wire transfers, ensure that detailed information is obtained on the sender and the beneficiary, and prohibit transactions by certain people pursuant to UN Security Council resolutions, such as Resolution 1267 of 1999 and Resolution 1373 of 2001, for the prevention and suppression of terrorism and its financing (Recommendation No. 16).
- Designated DNFBPs, such as casinos, real estate offices, dealers in precious metals or stones, and even attorneys, notaries, and accountants, must report suspicious operations, and those who report suspicious activity must be protected from civil and criminal liability (Recommendation No. 22, in combination with Nos. 18 through 21).
- Countries should take measures to ensure transparency and obtain reliable and timely information on the beneficial ownership and control of legal persons (Recommendation No. 24), including information on trusts—settlors, trustees, and beneficiaries (Recommendation No. 25).

Thus, it should be mandatory to record the true beneficial owners of companies, to keep records up-to-date, and to make records readily available to investigators in cases of suspected wrongdoing. The costs of openness will be outweighed by the benefits of shining light on the shady corners of finance. There is much intelligence

---

get away with. The City of London, which pioneered offshore currency trading in the 1950s, still specialises in helping non-residents get around the rules. British shell companies and limited-liability partnerships regularly crop up in criminal cases. London is no better than the Cayman Islands when it comes to controls against money laundering. Other European Union countries are global hubs for a different sort of tax avoidance: companies divert profits to brass-plate subsidiaries in low-tax Luxembourg, Ireland and the Netherlands." In *The Missing $20 Trillion. How to Stop Companies and People Dodging Tax, in Delaware as well as Grand Cayman. The Economist*, Feb. 16, 2013, http://www.economist.com/news/leaders/21571873-how-stop-companies-and-people-dodging-tax-delaware-well-grand-cayman-missing-20.

work to be done, more than that involved in simply controlling one's borders. Intelligence forces need to work together because if they are kept apart, each may, in isolation, feel that someone else is responsible for the problem.

Transparency will also help curb the more aggressive forms of corporate tax avoidance. The advantage of offshore accounts is that they enable the free movement of capital, which is taxed only in negotiations taking place in-country, with exemptions for transfers to other offshore or nonresident accounts, corporate income taxes, and income tax withholding on payments made to nonresidents. Arnaldo Sampaio de Moraes Godoy has apropos observations on these, especially concerning Barbados, Panama, the Bahamas, and Vanuatu.[3] Moreover, there are treaties to avoid double taxation that allow governments to establish unilateral measures domestically (such as, for instance, exemptions for fiscal credits at a reduced proportional rate and deduction of taxes paid abroad from domestic taxable income), which is why they are referred to as tax havens. The author cited also adduces that there is a draft being circulated by the Organization for Economic Cooperation and Development (OECD), called the *Model Tax Convention on Income and on Capital* (comprising of 31 articles distributed over seven sections), aimed at eliminating obstacles relating to double taxation. It provides, for example, that dividends paid by a company having its home office in one state party to someone living in another state party to the convention will be taxed by the latter, and their remuneration taxed by the state in which the services are rendered.[4]

It is true that offshore accounts facilitate the circulation of goods, services, and capital, but they are also an effective instrument for evading taxes with considerable legitimacy. They lend themselves not only to legal uses of arguable utility, but also to illegal practices. There are considerable advantages to be had by using them as conduits, especially by those interested in laundering ill-gotten money, on account of defective or nonexistent government control, and also because they make it easy to generate false trails and international wire transfers. Offshore bank accounts make it possible to disguise their real controllers, since ownership is—according to the legislation in the countries in which they are located—evidenced by bearer paper, and partners or officers are simply proxies, often proxies for hundreds of companies of the same pattern. All of this amounts to creating a veil for the actual owners to hide behind. Their paper cannot be traded on the domestic market, nor cashed in without considerable expense and questions about possible complicity in money laundering directed at anyone who converts it.

It is argued that offshore accounts are advantageous in that owning one does not involve liability to taxes, unless one were to actually invest in the country. Loan agreements are often written so as to lay hold of funds from offshore accounts without exposing them to tax liability. There are transparency requirements for beneficiaries of companies, with countries required to obtain reliable real-time information (FATF Recommendation No. 24), including information on trusts, settlors, and

---

[3] In *Direito tributário comparado e tratados internacionais fiscais*. Porto Alegre: Sergio Antonio Fabris, 2005, pp. 83–84.

[4] Cf. *id.* pp. 166–70.

trustees or beneficiaries (Recommendation No. 25), which would preclude anonymous accounts. This is why the customer and actual beneficiary must be identified (i.e., know-your-customer duties, often called CDD) along with a requirement to collect enough information about the institution to which service is rendered, so that the trustee, who administers the assets, is accountable for turning in suspicious transaction reports. Observe that the FATF takes a clear stand against the invocation of banking secrecy or professional privilege as a means of obstructing its recommendations (Recommendation No. 9).

04. Require tax havens to comply with all provisions whereby information must be provided to proper international authorities. This amounts to placing ethical and legal considerations above financial considerations, all the way down to obtaining information about the ownership (beneficial or ultimate beneficial ownership) and identifying the controllers. Indeed, the controllers merit special attention, and their very existence ought to be looked into to check whether they might be providing services to criminal enterprises. The hurdles in the way of their suppression are closely related to the Janus-faced discourse of many states which rely on tax havens to conduct nontransparent transactions purportedly having connections to "reasons of state" or for the management of assets belonging to their political elite.

9.1.1.4   International Legal Cooperation, Repatriation, and Extraterritoriality: Conflict of Laws

Explanation International cooperation must be improved to lend substance to the administration of justice and provide repatriation of assets through the following measures and national policies, mindful that the fight against crime is independent of where the crime occurred, and that confiscation is essential. As already established by the FATF, there is a need to make it possible to bring about freezing and seizure, even if the antecedent crime occurred in another jurisdiction (country). There is also a need for deployment of specialized multidisciplinary teams, task forces if you will (Recommendation No. 30). It also recommends international legal cooperation, pursuant to the UN Conventions of Vienna (international traffic, 1988), Palermo (transnational organized crime, 2000), and Mérida (corruption, 2003), by withdrawal of obstacles (Recommendation No. 37) and direct mutual assistance toward a quick, constructive, and effective solution (Recommendation No. 38).

The fight against organized crime must not be defeated by lack of understanding regarding the various international legal systems. Consider the impact on US asset forfeiture legislation of the case in which Brazil filed for freezing of assets belonging to a Brazilian defendant, and for keeping those assets in the USA. At issue was whether, based on 28 U.S.C. § 2467 (d)(3), called Enforcement of Foreign Judgment,[5] foreign assets may only be frozen after a foreign court has definitively

---

[5] *(d) Entry and Enforcement of Judgment—(1) In general—The district court shall enter such orders as may be necessary to enforce the judgment on behalf of the foreign nation unless the court finds that—(A) the judgment was rendered under a system that provides tribunals or procedures*

ruled in favor of forfeiture, or if it may be done before any final decision on confiscation has been rendered. The US Court of Appeals for the District of Columbia Circuit, on review of two decisions by the court in March and April 2009, decided that a final decision by Brazil regarding confiscation was required, according to its interpretation of 28 U.S.C. § 2467 (d)(3).[6] Following this decision, the US Department of Justice requested and obtained from Congress a resolution of the problem because, if upheld and followed, the decision would have compromised cooperation efforts with other countries.

Regarding conflict of laws and extraterritoriality, issues can arise when different legal jurisdictions disagree on the legality of the acts concerned. For instance, some countries criminalize online gambling activities while such activities are legal in other countries.

05. Cooperation through letters rogatory is not recommended because it is slow and bureaucratic, and because analysis in the requested country is limited to checks on public policy and affronts to sovereignty.

06. Prioritize cooperation by direct assistance as the response followed by states because it is faster, based on mutual trust, and conveys to the requested state a proper analysis of the requests.

07. Give preference, if possible, to the clear, simplified standard mutual legal assistance treaty (MLAT) format. Specific, separate MLATs should not be required for each asset, security, or pecuniary amount if the requesting state attaches to its request a list of assets and gives grounds.

08. Consider that central authorities have facilitated matters, for they placed no obstacles in the way of direct contact between magistrates or competent authorities, and channels of communication must be opened up to ease unnecessary bureaucratic burdens (Article 18.13 of the UN Convention Against Organized Crime at Palermo does not prohibit such understandings).

09. The regular legal systems of countries involved must be respected (requesting and requested states), and it is no bar to cooperation if the request originated with or was addressed to the police, the office of the public prosecutor, or the courts.

10. If extradition is refused, on the grounds of citizenship, then the person believed to be involved ought to be promptly submitted over to authorities in their own country (Art. 16.10, Palermo). However, if accepted, it is recommended that the sentence be served out in the requested state (Art. 16.11, Palermo), otherwise,

---

incompatible with the requirements of due process of law; (B) the foreign court lacked personal jurisdiction over the defendant; (C) the foreign court lacked jurisdiction over the subject matter; (D) the foreign nation did not take steps, in accordance with the principles of due process, to give notice of the proceedings to a person with an interest in the property of the proceedings in sufficient time to enable him or her to defend; or (E) the judgment was obtained by fraud. http://www.law.cornell.edu/uscode/text/28/2467.

[6] United States v. Opportunity Fund and Tiger Eye Investments, Ltd. No. 1:08-mc-0087-JDB, United States District Court for the District of Columbia. Decided July 16, 2012.   http://www.cadc.uscourts.gov/internet/opinions.nsf/1B9DC0B1D05DB6D5852578070070EC9C/$file/095065-1255619.pdf. Accessed June 14, 2012.

require serving the sentence or part of it in the requesting state (Art. 16.12, Palermo).

11. Consider the possibility of joint prosecution or transfer of criminal proceedings (Art. 21, Palermo) for final disposal of assets and joint measures (cooperative debriefings with effects in both countries) to achieve better administration of justice.

12. International cooperation ought not to be blocked while the whereabouts of an asset are unknown. The requested state should try all available means for tracing or seizure for future confiscation or repatriation.

13. As a condition for restitution, the requested state should require proof of the legality of the asset, security, or pecuniary amount whenever the requesting state requests for seizure with an eye to confiscation or repatriation, mooted by court decisions that did not rule on the merits as to the illegality of its origin.

14. Invocation of absence of dual criminality cannot justify failure to cite or subpoena defendants, victims, witnesses, or affected third parties once criminal proceedings have been initiated in the requesting state.

15. Information gained for criminal proceedings may be used in other such proceedings if the requested state so authorizes, even if retroactively.

16. Assets, securities, or pecuniary amounts shall be restituted for indemnification of victims or turned over to the United Nations Fund for technical assistance among countries—or even for reimbursement of the state. A division might be arrived at to deduct only expenses, except for such crimes as corruption and the like, and also with regard to cultural goods, which should be so disposed of as to give priority to public access.

17. Reimbursement of states should lie outside the reach of the statute of limitations, which does not affect international cooperation.

18. The denial of request of a court order for a mere citation, subpoena, or copies undermines international cooperation. It is incumbent upon states to simplify their legal systems to make direct assistance workable.

19. Defense witnesses ought to be heard in the country filing charges, or else by teleconference from embassies or consulates, with international cooperation not being invoked except in cases of evidence being disguised by the charges.

20. International cooperation does not require the attachment of proof, but rather, a presentation of arguments leading to the decision to see that measures be taken abroad.

21. Give weight to the primary bases of jurisdiction of nationality and territoriality to consider a violation of a crime adopting the method called "the test of reasonableness," which weighs the links between a state and its conduct against the links between its conduct and another state, allowing some flexibility.[7]

---

[7] BANA, Anurag. Online Gambling: An Appreciation of Legal Issues. 12 Bus. L. Int'l 335, 2011, p. 342.

9.1.2 A National Perspective

9.1.2.1 Institutional Measures (Executive and/or Legislative Branch)

Explanation The hazards of globalization may be minimized if with globalization, our notions of law draw authority from social and philosophical—as opposed to just economic—considerations. The subject cannot be dealt with from a purely economic view. Criminal organizations must be throttled by denying them what gives them their mobility and power, continuous and unprecedented illegal wealth. The subjective collective degradation afoot in the world today, which regards economics as the standard of value, can never so bind our numbers together as to gloss over such indispensable critical thinking. Legitimate social movements and individuals ought to assume a stance for ethical values to compel obedience to basic rules of coexistence. Standing these rules at defiance by parallel paths amounts to the real breakdown of rights both *de facto* and *de jure*.

There is a duty to perform CDD for all financial and nonfinancial activity, whether with natural or artificial persons, to ban anonymous accounts or those bearing fictitious names, and require identification of their beneficial owners, with records to be kept for at least 5 years (FATF Recommendation Nos. 10 and 11). There are, at times, requirements to make suspicious transaction reports on nonfinancial companies engaged in domestic or international cash transfer services, obliging them to record the amounts transferred, form of payment, transaction date, purpose of the wire transfer, name, individual or corporate taxpayer ID, where applicable, of both sender and receiver and addresses for both.[8] These requirements give a false impression that any money laundering occurring in that sector could actually be detected. There is also a need to make dealers liable to these rules, inasmuch as they may have knowledge or probable knowledge of criminal behavior (willful blindness doctrine).

9.1.2.2 Freezing, Seizing, Confiscating, and Repatriating Assets

22. Allow administrative freezing and seizure to be accomplished quickly, so as to prevent disappearance or acts of terrorism.

23. Allow confiscation of assets when transferred to an outside party which may have been aware that the assets were illegal, or that they were transferred solely to avoid confiscation.

24. Allow confiscation of illegal assets when a conviction cannot be obtained on account of death, statutory limitations, or grant of immunity. Adopt civil actions to terminate ownership.

25. Even after a decision has become final, allow further financial investigations to enforce prior confiscation orders covering all of the proceeds of crime.

---

[8] See, for example, the Resolution by the Brazilian financial intelligence unit, the Council for Financial Activities Control, COAF Resolution No. 10 of November 19, 2001.

26. Once criminal proceedings are instituted, the statute of limitations ought to stop running, for there is no reason to count government inertia, nor lack of interest in criminal prosecution.

9.1.2.3   Payments (in Cash/E-money, Stored Value Cards, Remittance companies, electronic payments, and via the Internet)

27. Require of all confederations, federations, and clubs—as ought to be required of all financial institutions—detailed records of loans, profit sharing, or other benefits to avoid substantial losses. Equity holdings acquired to accommodate subsidiary institutions are often not handled the same way as for institutions outside of the holding company. Note, for example, the absence of any formal equity holding or loan agreement, that is, the absence of documentation or frequent and excessive granting of benefits to partners.

28. Cash payments are practically untraceable and usually are the result of some sort of tax evasion or illegal act. Cash payments should therefore be banned for purchases of vehicles, boats, airplanes, real property, shares of stock, lottery tickets, bets in excess of US$ 10,000, and any gambling and sport transactions (player's transfers, media contracts, etc.). One thus tries to cut off the illegal flow of money and its entry into the legal market. This shifts us away from dereliction with regard to measures to curb money laundering and the financing of terrorism.

29. Payments by third parties should also be illegal, so as to preclude their use for purposes of masking real ownership of the goods and resources, with the potential for tax fraud that that entails.

30. Credit card payments (for example, for Internet betting) should also be subject to stricter controls. Lax controls have been exercised by card issuers, often with no prior relationship, exacerbated by failure to disclose credit limits, changes of address, name, date of issue, and expiration date. "Know your client" or due diligence rules must be enforced.

31. Ban all wire transfer payments that do not allow the money to be tracked. There is always a separation between the nonfinancial remittance company and the financial institution receiving the investor's money whenever these come from individuals or companies unrelated to the negotiations (cash deals negotiated by factoring companies or companies having home offices in tax havens). Currency brokers or hawala systems are often resorted to, as are wire transfers from and to secret banks or banks in tax havens—often by people having no connection whatsoever to the institution receiving the resources. They either are not account holders or are unconnected with the account holder receiving the wire transfer, or even transfer the amount that is actually the sum of many small deposits. Allowing this sort of practice would amount to having no preventive measures whatsoever. In wire transfers, detailed information should be obtained on the sender, as well as the beneficiary, with monitoring made possible, and there should be an option to prohibit transactions by certain people pursuant to UN Security Council Resolutions 1269 of 1999 and 1373 of 2001, for the prevention and suppression of terrorism and its financing (FATF Recommendation No. 16).

32. Establish clearly defined categories for stored value cards so as to make it easier for government agencies to identify suspicious cards, given that it is not easy to distinguish between traditional debit cards and prepaid access cards. Identification should be demanded of their customers, so as to enable comparison of identities of those individuals and wanted criminals. A top-down compliance program is needed, which should include customer identification, storage of records, and reporting of suspicious operations. Prepaid stored value cards have, with their explosive growth, become a perfect tool for money laundering whereby illegal money may be moved, with no documentation, identification, suspicion, or seizure. They are poorly (or sloppily) regulated, which ensures their anonymity, even when purchased or reloaded. Their daily limits are the same as the face value. They offer advantages over physical transfers because they are easy to carry and may be sent through the mail, and hence even replace the cross-border transportation of cash. Regulations alone will not suffice if they are issued by nonfinancial institutions. Their issuers' obligations must therefore be clearly established. They ought to be classified as monetary instruments and, therefore, be subject to declaration when going through customs. These cards, then, possibly in combination with cash or other monetary instruments, once aggregating the US$ 10,000 limit, ought to be subject to customs declaration.

33. Tighten controls on remittance companies, to have real knowledge of situations that might allow clandestine wires or wires not subject to suspicious operation reporting requirements (poor or borderline tracking by authorities). One example would be requiring a declaration by a bank accredited by the Central Bank to handle the conversion whenever called upon by the national government to appear for settlement of the currency exchange operation. Detailed information should be obtained on the sender and beneficiary of the wire transfer, with monitoring made possible, and the option of prohibiting transactions by certain people pursuant to UN Security Council Resolutions 1269 of 1999 and 1373 of 2001, for the prevention and suppression of terrorism and its financing is FATF Recommendation No. 16.

34. There has to be some form of electronic tracking of payments over the Internet, which is possibly even easier than for cash money that simply passes from hand to hand. This would enable detection, for example, of payments made using electronic payments (for instance, bitcoins), even though considered a transparent method of conducting transactions inasmuch as the system allows identification of its users, albeit in aliases or nicknames. Hence, if a trafficker is using a given electronic payment address, all data on the person using that address and the entire graph of parties with whom that trafficker has dealings should be accessible. Although the medium is not as anonymous as would appear at first blush, there is no denying the possibility of someone setting up channels on the bitcoin system to conceal transactions behind anonymity. As more services like this come online, the more complex the transactions, and the greater the opportunity for apparently unrelated, off-the-grid transactions would be—especially when it is possible to use a number of different addresses, with each one used for only a single transfer.

9.1.2.4   Offshore Accounts and Trusts

Explanation According to Susan Ormand, "Some large banks and credit card companies, including Visa and Mastercard, have refused to transfer money to offshore betting accounts because of the potential for fraud, but bookmakers circumvented these efforts by encouraging bettors to use debit cards or payment services like PayPal."[9] Obtaining information has been hampered considerably by a lack of channels of communication with the competent authorities—to say nothing of timely notice—in the conduct of international legal cooperation. Moreover, investors may have committed antecedent crimes and wish to launder the proceeds.[10]

35. Full particulars absolutely must be obtained on all actual investors, even if they belong to companies chartered abroad, provided they do business and are represented within a given country. A simple listing of proxies or stockholders is not enough. Complete identification must also be required of partners and administrators concealed within offshore accounts or trusts domiciled in tax havens. A listing of all partners and administrators ought to be required for being listed or removed from the tax rolls (which in Brazil is the Treasury Ministry's National Corporate Register or CNPJ).

9.1.2.5   NGOs, NPOs, and Foundations

Explanation There ought to be a complete record by type of business and types of NGOs. NGOs should be required to keep records on all transactions entered into within the country and/or abroad. This would comply with FATF Recommendation No. 08, in the spirit of clearly delimiting the rights and responsibilities of directors and employees of NGOs. It would encourage countries to establish good policy whereby information on their activities, size, and other important characteristics such as transparency, integrity, openness, and best practices can be had in real time for purposes of supervision and monitoring (FATF Recommendation No. 8). There is evidence that such organizations have even been used for the financing of terrorism. Government policies have been initiated to curb the practice, most notably in

---

[9] In *Pending U.S. Legislation to Prohibit Offshore Internet Gambling May Proliferate Money Laundering*, 10 Law & Bus. Rev. Am. 447, 2004, p. 451.

[10] In Brazil, the Internal Revenue Service Regulatory Directive No. 748/RFB of June 28, 2007, does not require offshore accounts or trusts to completely identify stockholders and administrators as is done for companies domiciled in Brazil. All that is required for enrollment in the Corporate Taxpayer Register (CNPJ) is a copy of the company charter. This means they might be satisfied with a simple declaration issued by a tax haven's public agency (bearing the company name, date chartered, type, purpose, and address).

Pakistan.[11] Cases of NGO involvement in terrorism and its financing have also been reported in India.[12]

36. Licensing should be required for operation under tax exempt status, and continuation of this status ought to be contingent upon regular reporting to revenue authorities of all relevant information, in an official document duly dated and signed under penalty of perjury, listing the name and telephone number of the person in charge of the books and records of the organization or foundation.

37. The organization's books and records ought to include a detailed list of its activities and management, all revenue and expenses, and liquid assets, to include: the name and purpose of the institution; number of members; whether they have on hand more than 25% of their liquid assets; number of voting members listed within and outside of the organization; number of employees; number of volunteers; unrelated business revenue and amount paid in taxes; contributions and donations; resources invested; benefits paid to and for members; total assets and liabilities; basic description of all assistance programs; whether any loans or benefits were granted to employees, directors, a trustee, or any other person; name, number of hours worked, and job description of all employees and former employees (including directors, trustees, and key personnel); earnings had by these individuals; expenses claimed (including travel and entertainment); and names and particulars of all donors.

38. An external audit should be required above a given gross revenue ceiling (more than US$ 100,000, for example), as the state of New York so capably provides.[13]

39. They should include in their bylaws requirements for distribution of financial reports and outside audit reports for all directors and management personnel (president, manager, and financial department), for easy review.

40. Require universal, electronic access to the records, and required reporting documents of all NGOs, associations, and foundations.

41. In the case of a temple, church, mosque, educational institution, or trust, even if registered as an NGO, association, or foundation, all sources of funding must be provided in sufficient detail.

42. There should be a bar on receiving cash donations, or at least a cap above a certain amount (say, US$ 3,000), which would, above that amount, restrict donations to banking instruments.

43. Accounts of all such entities should be reviewed to reinforce due diligence and check whether they actually perform the purposes for which they are organized.

---

[11] Cf. Terror outfit-turned "charity" JuD set to come under Pak Central Bank scanner. In *Asian News International*,03/13/2012. www.lexis.com. Accessed June 19, 2012; Naveed Butt. Insurance/takaful companies: SECP enforces compliance with AML Act. *The Financial Times Limited.* NPO 3/25/2012, 2012 WLNR 6308356.

[12] See Prafulla Marpakwar. State forms cells to detect source of terror funds. *Times of India.* Copyright 2011 Bennett, Coleman & Co. Ltd. December 24, 2011. www.westlaw.com. Accessed June 19, 2012.

[13] Cf. www.charitiesnys.com, or www.chaririesnys.com/pdfs/statute_booklet.pdf. Accessed May 29, 2012.

Accounts should be allowed to be opened only in the organization's own name and in accordance with the documentation turned in.

44. If announcements are made that a given account will be receiving donations or something similar, banking institutions must monitor this to check on the beneficiary of wire transfers made from that account, and promptly make out a suspicious activity report to the FIU if the published account is different from the account owned by the NGO or foundation.

45. Check that all donations and contributions received for specific purposes are being properly recorded and faithfully accounted for.

46. Provide clear procedures for board membership, thus ensuring diversity among the members.

47. See to it that all board members act in good faith to avoid any conflict of interest between the entity, its purposes, and themselves.

48. Secure independent and exempt financial evaluation.

9.1.2.6   Money Laundering as a Crime

Explanation The communications system for reporting suspicious transactions is the key to effective suppression of money laundering. It turns up a number of suspicious deals. Many others go unnoticed when there is no cooperation from those whose legal duty it is to report transactions. Certainly the failure of one of the methods of control held to be essential in the fight against money laundering, namely, reporting, can give rise to misleading statistics. Moreover, making one liable to criminal charges for incorrect notices of suspicious transactions is clearly aimed at protecting privacy and image, on the one hand, and the effectiveness of early investigations on the other, for the danger is that future freezes on accounts and other confidential security measures might be rendered inoperative. The FATF noted three characteristics of the Internet that make it potentially susceptible to money laundering: ease of access, depersonalization of contact between the customer and the institution, and the rapidity of electronic transactions.[14]

49. People who do the following should be liable to criminal prosecution: (1) fail to report suspicious operations, (2) delay in reporting suspicious operations, (3) make incomplete or false reporting, (4) make public the required reporting, or (5) structure transactions or operations to circumvent reporting requirements.

50. Require the preparation of suspicious activity reports by individuals, companies, casinos, lottery houses, confederations, federations, clubs that sell, import, export, or intermediate a transaction—whether on a permanent or temporary basis, in a principal or accessory role, and cumulatively or otherwise, to prevent the laundering of money through gambling and sport.

---

[14] FATF, Vulnerabilities of Casinos and Gaming Sector, March 2009, available at www.fatf-gafi.org/dataoecd/47/49/42458373.pdf.

9.1.2.7   The Administration of Justice and the Role of Judges

Explanation The numerous complex actions involved demand considerable specific knowledge and may entail huge setbacks to criminal jurisdiction for failure to act at the proper time.

51. Have courts specialized in money laundering, with criteria established for recommending, appointing, and replacing judges, in addition to specialized criminal teams at the courthouses.

Explanation Reintegration of financial criminals into society must be centered on making them rethink their behavior. If there is indeed any reasoning behind unlawful conduct involving cost–benefit analyses of the outcomes to the perpetrator, a given crime will be committed if and only if the expected penalty is outweighed by the advantages to be had from committing the act.[15]

52. Pecuniary penalties ought not be so freely applied, and should perhaps even be restricted, and the same goes for incarceration (even in plea bargaining), when there is a clear lack of intimidating effect. The requirements of proportionality (gravity of the crime plus guilt) and the need for overall prevention require a response that is a better fit for serious financial crimes.

9.2   Specific Proposals—Gambling and Sport/Football

9.2.1   Regulatory Agencies

Explanation States must be able to identify, disrupt, and dismantle networks that engage in money laundering and terrorist financing. The international FATF compels states to fight money laundering through various fronts, including an aggressive policy to allow its discovery and eradication.

Gambling A domestic prohibition of Internet gambling would not likely have an effect on Internet betting as a money-laundering platform because many jurisdictions have legalized Internet gambling. Thus, international cooperation on this issue is a difficult task. The legalization and regulation of Internet gambling would be a better solution to avoid money laundering. A well-regulated Internet casino would not be a good vehicle for money launderers because all gambling transactions could be recorded and readily traceable. According to Mark Schopper, "credit cards, checks, or money transfers could be used at the sites removing consumer need for electronic money, and creating an audit trail for law enforcement to monitor suspected criminals. In addition, even if e-money was used at domestic Internet

---

[15] For more on this, see Jesús-María Silva Sánchez (in *Eficiência e direito penal.*Coleção Estudos de Direito Penal. São Paulo: Manole, 2004, No. 11, p. 11) and Anabela Miranda Rodrigues (in Contributo para a fundamentação de um discurso punitivo em matéria fiscal. *Direito Penal Económico e Europeu: textos doutrinários.*Coimbra: Coimbra ed., 1999, pp. 484–85).

casinos, other verification processes could be developed to identify the gambler. Indeed the best way to monitor suspicious currency transactions is to 'monitor' them."[16] The registration services and Internet service providers require reliable billing information in exchange for their services. According to Jon Mills, "there exists a compelling state interest in regulating the gambling industry. This interest does not diminish when the gambling occurs over the Internet; in fact, the government's concerns are magnified." [17] According to John Warren Kindt and Stephen W. Joy, "countries combining lax Internet casino regulations with substantial privacy laws created a recipe for disaster."[18]

On the other hand, Internet service providers can join a regulatory force on the Internet by blocking objectionable sites, thus forcing bettors to gamble through unreliable offshore operators.[19] The attempts of the US Congress to prohibit Internet gambling, contained in proposed legislation, the Unlawful Internet Gambling Funding Prohibition Act (UIGFPA), would institute a monetary control scheme preventing the use of credit cards and other bank instruments (checks and electronic funds) in Internet gambling. The government, according to Mark D. Schopper, "develops what appears to be an 'effective' method of achieving this goal: a monetary control scheme that would effectively make it impossible for Americans to fund their Internet gambling activities."[20] These measures are being considered by the government to thwart criminals and terrorists from laundering the proceeds of their illegal activity through Internet gambling sites. However, due its popularity, it is not easy for the government to eliminate it without great difficulties since alternative payment schemes would be available to fill the gap left by credit cards. E-money, for example, is based on technology that makes enforcement of a ban impracticable at this time.

Sport FIFA, confederations, federations, clubs, and associations should join government enforcement agencies in the effort to curb money laundering. The attempt of FIFA to obtain vital information through the Transfer Matching System (TMS) is valid but not enough. It is a vital tool for obtaining information about the international transfer of players, previously restricted to only business stakeholders. Through this system, more than 30 pieces of information are recorded online, such as player history, clubs involved in the business, payments, values, contracts, and

---

[16] SCHOPPER, Mark D. Internet Gambling, Electronic Cash & Money Laundering: The Unintended Consequences of a Monetary Control Scheme. 5 Chap. L. Rev. 303, 2002, p. 328.

[17] In Internet Casinos: A Sure Bet for Money Laundering. 19 Dick. J. Int'l L. 77, 2000–2001, p. 108.

[18] In Internet Gambling and the Destabilization of National and International Economies. Time for a Comprehensive ban on Gambling Over the World Wide Web. 80 Denv. U. L. Rev. 111, 2002–2003, p. 119.

[19] Ian Abovitz. Why the United States Should Rethink Its Legal Approach to Internet Gambling: A Comparative Analysis of Regulatory Models that Have Been Successfully Implemented in Foreign Jurisdictions. 22 Temp. Int'l & comp. L. J. 437, 2008, p. 468.

[20] In Internet Gambling, Electronic Cash & Money Laundering: The Unintended Consequences of a Monetary Control Scheme. 5 Chap. L. Rev. 303 2002, p. 304.

other kinds of information. It is possible to check the information received from both the buyer club and the selling club. For instance, one can check to see whether the contracted amounts were allocated directly to the involved parties before such amounts have been reported to the recipient bank accounts. This is very important, especially when the information is electronically available. But FIFA should not be the only recipient of such data, as the information would be very useful to authorities investigating money-laundering crimes.[21]

It is essential to require clubs, federations, and confederations, as well as those who provide advisory, auditing, bookkeeping, and consulting in this area, to communicate suspicious transactions to the FIUs. There are records of money laundering with the involvement of clubs in the negotiation of international money transfers in various countries. Clubs, according to the FATF, are deliberately being used to launder money. FIFA, an organization that sometimes acts purely with commercial and private interests, should not replace authorities. FIFA data are not public and not easy to obtain, and authorities would be forced to request international legal cooperation to access them because FIFA is headquartered abroad. It is important to set up a supervisory body to be able to closely monitor the industry, including the management of clubs that often bear debts incompatible with the effective financial capacity. That is why it is insufficient to create a system where information is only supplied by these clubs without any control over them.

The creation of an agency similar to the National Directorate of Management Control (DNCG) of France would signal strong repudiation of the unlawful practice of money laundering. The agency should be composed of people from the sport sector as well as members of the federal police, prosecutors, etc. It should be fully independent and have specific powers regarding the actions of sports clubs.

Large amounts of money should not be moved in the sport sector without adequate supervision and regulation. More effective action by the authorities to address the issue at the national level is essential in order to provide the necessary transparency in financial operations connected with football and the prevention of money laundering and tax evasion. The poor management and fraudulent administration of clubs must be detected and punished. The regulator should have the power to ban negotiations that could jeopardize the healthy finance of clubs, federations, and confederations.

The requirement of financial transparency of professional football clubs would materialize through the publication of their accounts, which would gather financial information on all professional clubs in the country. Corruption frequently affects the sport sector given its strong internationalization. As the sport sector cannot solve the problem alone, sport organizations need to work more closely with the government. The development of public–private partnerships at the national and international levels is considered an important component in the fight against problems such as corruption, money laundering, and match fixing. Distrust of football leaders

---

[21] FIFA denied Spanish Courts to get a copy of the contract between the Brazilian football player Neymar and Barcelona club (see http://esporte.uol.com.br/futebol/ultimas-noticias/2014/02/05/fifa-se-nega-a-enviar-contrato-de-neymar-para-justica-da-espanha.htm. Accessed Feb. 06, 2014).

has reached an unsustainable level and undermines the credibility and dignity of sport competitions. The threat of suspension by FIFA when countries have no intention of investigating draws attention and serves to alert the authorities.[22]

53. Establish a regulatory agency, if it is not be feasible for the FIUs to provide proper oversight so as to timely obtain direct or indirect access to financial and administrative information, and information from law enforcement authorities in order to fully perform their functions, including analysis of suspicious transaction reports (thereby fully complying with FATF Recommendation Nos. 26, 27, 29, and 31).

54. The regulatory agency shall be empowered to demand secure records with profound and specific evaluations of individuals (photo ID, proof of domicile, employment records), and of similar institutions for constant review, physical whenever possible—for many entities stop everything whenever they believe that companies similar to their own are having to enforce compliance.

55. Compare records with the contents of other databases to determine whether the underwriters demanded complete documentation before the transaction. Whenever some third party is involved, check whether guarantees are accepted without proper appraisal.

9.2.1.1 Violation of Consumers' Rights or Antitrust Laws and Underage Internet Betting

Explanation Breach of trust violates consumer rights and compromises fair competition in the market. Good faith or loyalty contracts should be a rule of conduct submitting to administrative penalty in order to prevent fraudulent business practices. The misconduct is not acceptable.[23] On the other hand, it is important to provide a minimum level of protection for all consumers regardless of their host country's position. Consumer protection and interest must ensure that gaming and sport are conducted in a fair and open way. According to Charles E. Brown, "antitrust laws are designed to maximize consumer welfare by minimizing excessively large combinations of economic power."[24]

56. Require punishment of business practices that violate good faith or the loyalty contract that are designed to protect by consumers' rights. They are supervised by, in Brazil, the National Consumer Protection System, and in the USA, the Federal Trade Commission and the US Department of Justice.

---

[22] Andrew Jennings, *in*: A ginga perfeita dos donos da bola. A FIFA controla o dinheiro, marca os adversários e dribla a Justiça. Entrevista. *O Estado de São Paulo*, J4, *aliás*, June 27, 2010.

[23] See Humberto Theodoro Júnior. *Consumer Rights* (*Direitos do Consumidor: a busca de um ponto de equilíbrio entre as garantias do Código de Defesa do Consumidor e os princípios gerais do direito civil e do direito processual civil*). Rio de Janeiro: Forense, 2009, p. 25.

[24] In Professional Football and the Antitrust Laws: Impact of *United States Football League V. National Football League* and a Proposal for Change. 31 St. Louis U.L.J. 1057, 1986–1987, p. 1058.

Explanation Most jurisdictions stipulate 18 as the age of majority for online gambling or online betting sites. The focus on regulation of those activities should prevent these sites from reaching minors. From a social perspective, the accessibility of gambling on the Internet placed an addictive behavior in the hands of millions of previously unexposed people, including adolescents and impressionable children.[25]

57. Enact laws to prohibit unwanted advertisements in jurisdictions that ban gambling for people under 18 years of age.

58. Require that operators verify a patron's credit card information with reliable third-party systems and with physical copies of a driver's license or a utility bill on registering.

9.3   Specific Proposals—Gambling

Explanation Casinos, lottery houses, and online gambling have been used for money laundering since people may falsely identify themselves or use a front person or company, even when withdrawing their prizes. Because the Internet is open 24 hours a day, 365 days a year, an individual with Internet access can place an online bet whenever he or she wishes. According to Michael Blankenship, "In a government-sponsored report on the impact of Internet gambling, the federal government claimed that 'the high-speed instant gratification of Internet games and the high level of privacy they offer may exacerbate problem and pathological gambling'."[26] Therefore, strict control of Internet gambling sites, subjecting them to auditing, is justified.

59. Tax the Internet gambling industry. Tax the gambling winnings of all gamblers and permit loss deductions for professional gamblers.

60. Require casinos, lottery houses, and online gambling sites to provide complete and detailed information to the Internal Revenue Service (IRS) or when demanded per customer (name, address, identity, activity, fingerprint, gains or losses arising, games played), under penalty of punishment. The same must be provided to FIUs.

61. Require annual gross revenue to be reported when it exceeds a certain value (US$ 1,000,000.00) and limit individual customer transactions per day (US$ 10,000.00).

62. Monitor customer activities above a certain threshold (US$ 3,000), regardless of who the customer is (including special clients).

63. As for the awards, require delivery only to the real winner. For alleged overseas earnings, require a detailed statement from the casino or lottery house.

---

[25] Cf. JOY, Stephen; WARREN KINDT, John. Internet Gambling and the Destabilization of National and International Economies. Time for a Comprehensive ban on Gambling Over the World Wide Web. 80 Denv. U. L. Rev. 111, 2002–2003, p. 126.

[26] In The Unlawful Internet Gambling Enforcement Act: A Bad Gambling Act? You Betcha! 60 Rutgers L. Rev., p. 501.

64. Prohibit the ability to acquire casino credit that is not intended to be used for gambling. Prohibit the exchange of money for money, by check or wire transfer, and even the simple exchange of money for credit by paying or not with debit or credit cards.

65. Require the collection of complete identification information from those who exchange for cash the credit remaining in their casino accounts after gambling.

66. Grant only nontransferable and personalized credit cards to casino customers.

67. Establish an expiration date for remaining credits or chips that cannot be exchanged in outside houses, and even when exchanged in the very house in which were played, are allowed only to the person who acquired them.

68. Ban the negotiation of chips or credits in the market.

69. Financial institutions should give greater attention to accounts in the names of politically exposed persons in casinos and lottery houses.

70. Special attention should be given to the deposits in current accounts of gambling houses by filling in specific detailed forms with the activity performed.

71. Credit purchases above a certain threshold should be automatically reported because of the suspicion that such a purchase is connected to money laundering.

72. Permit "junket" activities (i.e., excursions providing tourists access to casinos), even when on ship casinos, only when they are registered with the competent surveillance authority and fingerprints of those involved are provided.

73. Properly train employees of the gambling houses.

74. Provide detailed information to customers about safe boxes or special services.

75. Award managers that meet regulatory obligations to prevent money laundering.

76. Create, when necessary, special police stations tasked specifically with money laundering investigation.

77. Require regular audits, including slot machines.

78. Make it an obligation for lotteries to name and identify lottery ticket winners (profession, identification and IRS numbers, home address, and fingerprint).

79. Require written justification for high gains that practically defy chance.

80. Create of a national registry of players, divided by place of practice, age, winnings, and identification.

81. Interact with the international counterpart through bilateral or multilateral channels of communications between agency bodies in charge of regulating gambling activities.

82. Require individuals and companies that, on a permanent or occasional basis, as principal or accessory, together or separately, act in the promotion, brokerage, marketing, or trading of securities arising from gambling done in casinos, gambling houses, bingos, lotteries, or the like, to report suspicious activities.

Explanation If a rigid accountability system is established, then authorities will be able to track cash transfers and currency exchanges.

83. Establish strict control and supervision by an agency equipped with human and material resources. Controls should include accounting and audit opinions, and asking for full identification of customers and owners of lottery houses.

9.4   Specific Proposals—Sport/Football

Explanation The European Commission released a document urging member countries to create the crime of sporting fraud. It also recommended creating uniform controls, harmonizing rules in different countries, and empowering players to resist any pressures. Players should not have incentives to resort to organized crime. Illicit trade in sport can take place concerning not only the outcome of a game but also the number of points or goals, the amount of corners, or yellow or red cards, etc. To combat fraud, sport gambling sites and the flow of money must be controlled, as explained above.

84. Create the crime of sporting fraud.

85. The members of the FATF should consider determining how players' agents (including individuals or legal entities that promote, mediate, trade, hire, or negotiate athletes' transfer rights) are obliged to report suspicious operations.

86. Football clubs should be obliged to keep records of every contract and related mediation contracts for at least 5 years.

87. Foreign exchange contracts arising from remittances to individuals or legal entities related to football should be guaranteed by the contracts between the clubs and the football players.

88. FATF members should consider having the full identification of the investors, even when corporations in the country represent them.

89. More requisites of control and registry of the origin of the account holders and the beneficiaries of the money that is remitted to tax heavens should be applied. Further mechanisms should be designed in order to try to get tax havens to provide all kind of information, in a timely manner, when requested.

Explanation The absence of an obligation for those who act in player promotion, brokerage, marketing, and trading has allowed the practice of money laundering. According to Noël Pons, football clubs are businesses, and the phenomenon of money laundering should be given the same level of concern that for enterprises.[27]

90. Require individuals, corporations, associations, federations, confederations, and clubs that are involved in the promotion, brokerage, marketing, or trading of athletes to report suspicious transactions regarding negotiations.

Explanation Well-defined ethical and legal limits are necessary in order to educate everyone about the problem, notably the impossibility to invoke confidentiality to prevent the reporting of suspicious transactions.

91. The Federal Reserve, FIUs, the IRS, and the Securities Exchange Commission should offer training courses in business management to clubs and transfer agents in federations, confederations, and any other supervisory body, with the aim of strengthening their roles.

---

[27] Cf. Blanchiment d'argent: l'autre mercato. Le Monde, published August 8, 2012. In http://www.lemonde.fr/sport/article/2012/08/30/blanchiment-d-argent-l-autre-mercato_1751790_3242.html. Accessed May 6, 2013.

Explanation Organized fans must be properly monitored in order to avoid the creation of privileges to them (facilitated tickets, special meetings in clubs, transportation, police escort) because it has served as a stimulus to sublimate irrational expressions of collective actions. In addition, they must have the obligation to keep records with photos and addresses of all members to make it possible to punish the misconduct of an associate.[28]

92. Require organized supporters or fans to register with a regulatory agency and provide training to their managers, with the obligation to report suspicious transactions.

Explanation Any relaxation of the rule that aims to protect minors would facilitate criminal activity because it easily permits cheating.

93. Avoid loopholes that end up exposing those who were the object of special protection. For example, eliminate the exceptions to the rule protecting minors (such as the FIFA rule in Regulation on the Stature and Transfer of Players, Articles 19 and 19a).

Explanation Confederations and federations usually do not know how to inform relevant aspects of such transfers since they do not receive full information surrounding this business because their access is limited to the Certificates of Transfer.

94. Require clubs, federations, and confederations to comply, under penalty of sanctions, with the Registration of National or International Players Transfers. They must provide complete information of the transaction, by detailing its financial structuring; they must attach to the transfer agreement between buyers and sellers, the agent contract, and proof of identity of the agent and the player.

Explanation This will hamper fraud aimed at circumventing the ban on minors in professional sports.

95. Require clubs and training schools to register comprehensive information on each athlete and to forward this information to federations, confederations, and authorities when demanded.

Explanation People not connected to the sport are acquiring players' rights through offshore companies with complex and impenetrable financial structuring.

96. Require those who act as agents of athletes, even relatives or lawyers, to obtain a special license in order to avoid the nontransparency of their activities.

Explanation On the pretext of investing in football, a person can compromise clubs' management and losses of dubious origin are accepted.

97. Prohibit contracting services from companies owned by officers or persons related to sport by clubs, federations, or confederations in order to avoid conflicts of interests and illicit gains.

---

[28] See the Brazilian legislation, Estatuto do Torcedor, Law 10.671 of May 15, 2003 and Law 12.299 of July 27, 2010.

98. Regulate the legal framework of football agents to include all trading beyond the clubs.[29]

Explanation Under FIFA rules, an agent is a person who performs a paid function that takes a player to negotiate or renegotiate a contract of employment with a club or two clubs through a contract transfer. The concept is quite simple because the agent can perform a variety of functions other than the function of negotiating with clubs and athletes. For example, Felipe Legrazie Ezabella pontificates that there are agents who advise on several other matters and contracts and negotiate with several other companies, entities, and persons. For example, an agent prospects license agreements for image use, sponsorship, and advertising; advises in tax, labor, accounting, and investments; manages insurance premiums and private pension systems; coordinates travel; assists in personal and family matters; and chooses medical treatment, future career paths, staff, etc. Under FIFA rules, these activities cannot be performed by directors and coaches (Article 1.3 of the Players' Agents Regulations, which entered into force on March 1, 2001), which may result, in the absence of supervision, in damage to the sport and to many athlete victims because athletes may be deceived by false promises and fake agents who abandon them.

On the other hand, discussing FIFA's regulations, James G. Irving stated that "first, the regulations do not cover amicable transfers. Therefore, transfer of astronomical values will continue where the clubs and the player all agree on the transfer. Second, the regulations permit a player to unilaterally breach his contract and join another club (who is willing to buy out his previous contract) after only a four-month suspension. Some have voiced a concern that rather than the 'sporting sanction' it was intended to be, the 'four month rule' will become the default transfer rule. Additionally, with the 'sporting just cause' exception, the regulations provide an escape for a player from the sporting sanctions."[30]

99. Require players' agents to be licensed, so as to lend some transparency to their dealings.

100. Regulate and supervise all activities of players' agents, with required licensing or special authorization (personal and nontransferable), to the extent that they might represent interests more inclined toward the financial exploitation of sport and lack any commitment to its importance and credibility.

---

[29] On March 21, 2014, the FIFA Executive Committee approved the Regulations on Working With Intermediaries, proposing a new system that intends to be more transparent and simple. After April 1, 2015, players can choose any parties as intermediaries, but will have to respect certain minimum principles. Players will no longer have "agents," and all responsibilities will be transferred to players and clubs, which must act with due diligence when selecting an intermediary. A registration system for intermediaries will be put in place at member association level. See FIFA Executive Committee Approves Regulations on Working with Intermediaries, March 21, 2014, available at http://www.fifa.com/aboutfifa/organisation/administration/news/newsid=2301236/. Transferring all controls to clubs can, contrary to the intention of FIFA, stimulate money laundering activities in cases of collusion between clubs and players.

[30] IRVING, James G. Red Card: The Battle Over European Football's Transfer System. 56 U. Miami L. Ver. 667, 2001, p. 724.

Explanation The imposition of legal limits prevents deleterious effects upon the intermediation of athletes. The license, which is personal and nontransferable (Article 12), is sufficient to prove an unblemished record and is available upon written request to the national association. If the agent has more than one nationality, the application shall be to the country that the agent most recently acquired his nationality. Any person seeking to receive an agent license should never have been convicted in criminal cases, violent or financial (Article 6.1 of the Regulation), and should not have been charged with tax debts. Furthermore, the applicant should not perform any function in FIFA, confederations, national associations, clubs, or any organization linked to the latter (Articles 6.2 and 6.3). It would therefore be appropriate to require the applicant to present his curriculum to scrutinize the companies where he worked, job functions, study sites, references, and expertise.

To avoid conflicts of interest, all contractors should have the duty to inform their customers of the identity of their other clients. Customers then may end the contract at any time by invoking incompatibility of negotiation in the case. This will prevent situations in which an agent intermediates a deal for a club and for an athlete of the same club at the same time, for example. "Drawer" contracts that circumvent the rule that prohibits automatic renewal justify the demand for recognition of signatories that proves the correct date of the conclusion of the contract. It should clarify who is responsible for paying the agent (player or club) and the form of payment. The general rule is that the player pay for it, but it may be agreed in writing that the club is responsible for it (Article 19.4). However, the imposition of a limit (a percentage) on the gain of the athlete or club avoids excessive and disproportionate gains as commission for transfer agents (e.g., between 2 and 5 % per year). The legal representatives of minors cannot contract in their name for issues beyond the limits of mere administration of property, except by necessity or apparent interest, without the prior approval of a judge. So, the transfer of "economic rights" that is a future right requires permission under penalty of nullity.[31]

101. Establish legal limitations for doing business as a player's agent. Registration must be required, with a detailed résumé, in a regulatory agency, like the Federal Reserve or and the IRS, in addition to FIFA.

102. Impose a percentage limitation on profits.

103. Bar all those with criminal convictions or those who lost civil cases relating to fraud, tax evasion, or other civil liabilities, in state, municipal, or federal levels.

104. Prohibit the inclusion of irrevocability clauses in contracts of representation for children under the age of 18 and limit the percentage perceived as fees.

Explanation The Regulation on FIFA Players' Agents was initially adopted on May 20, 1994, amended in December 1995 and on October 29, 2007, and went into effect on January 1, 2008. It clarifies that an agent is an individual empowered, under license, to care and to act on behalf of players and/or clubs, about players' transfers anywhere in the world. On December 10, 2000, the FIFA Executive Committee approved a new regulation, Players' Agents Regulations, which entered into force on March 1, 2001, to fit the requirements of the European Parliament. It

---

[31] See Article 1691 of the Brazilian Civil Code.

allows any resident of the European Union or European Economic Area to request an agent license in any country where the national federation has its legal domicile. It determines, with the enactment of a new Regulation, dated October 29, 2007 and entered into force on January 1, 2008, that each national association prepare its own rules based on FIFA's rules. The FIFA Players' Committee must approve these rules. It establishes a limit on the duration of the FIFA's agent license (5 years) and subjects the license holder to further assessment when the license expires. No athlete or club is forced to use an agent, but if so it requires a license issued by national association (Article 3.1 of the Regulations of FIFA) to negotiate or renegotiate contracts with clubs, except in cases of strict confidence like a regularly enrolled lawyer in the country where the player lives or in the case of a sibling, spouse, or parent of the athlete. These hypotheses do not submit to the jurisdiction of FIFA. Article 1 of the Regulations makes it clear that the rules only refer to players' agents and clubs, so they are not applicable to any service provided to directors and coaches, who must submit to rules in the workplace. Mere performance of FIFA would not be enough to deal with the issue since it would not be able to inhibit the action of opportunists who give no contribution to the training of athletes because they do exclusively speculative investments with violation of ethical legal precepts.[32] On the other hand, the duty of obedience also to international standards, besides international legal permission to adhere as a voluntary membership by confederations and federations to the determinations of FIFA, can end up imposing a violation of national sovereignty. Only in the organization and functioning of the sport must FIFA's standards be applied.[33] Questions arise as to the legitimacy of specific regulations of FIFA as the pretext to better serve sport. For example, the obligation to license FIFA's agents only to individuals, banning clubs and coaches to use their services without a license (FIFA Players' Agents Regulations). Such agents stormed the market and have defended interests that often result in economic exploitations and several parallel businesses that gravitate around the sport industry.

105. Compel the agents to inform all customers as to their contractors. Prohibit brokering a club athlete and this same club simultaneously. Require recognition of the signatories of firms in contracts with agents. Prohibit the undertaking of obligations that exceed the limits of simple administration of goods, except for necessity or apparent interest, but with the prior approval of a judge in the case of contracts between agents and legal representatives of children and adolescents. Thus, the assignment of "economic rights," which is a future right, requires permission, under penalty of nullity.

Explanation Eliminate the possibility that professional athletes become prisoners of entrepreneurs who, in practice, with the end of the Pass, take possession of the athlete via proxies and irrevocability clauses. Before the performance of the player or a club, it imposes the formalization of a written contract (Article 19.1 of the FIFA

---

[32] Brazilian law, in § 7 of Article 28 of Pelé Law, addresses the issue of sports agents, only prohibiting the granting of proxies to them with a term exceeding 1 year.

[33] See Article 217, Part I, of the Brazilian Federal Constitution.

Regulation), with a maximum term of 2 years and is not subject to tacit renewal (Article 19.3). The prohibition or nullity of exclusivity clauses if the agent does not participate effectively in the agreement is due to the idea that there is no payment without work and no one can stop self-administration of personal interests.

106. Prohibit the automatic renewal of attorney letters in case of silence. The same goes for contracts with agents. Prohibit exclusivity clauses or make them void in cases of nonparticipation of an agent.

Explanation The rules set out the autonomy of sports bodies and association leaders regarding the sport's organization and operation. They still establish that the judiciary only admits cases regarding discipline and sports competitions after the exhaustion of the instances of the sports court (faster, with the administrative units and constituting independent and autonomous body). Sport is regulated by formal national and international standards and the rules of each sport mode have been accepted by the respective administration of sport. The fact that sport is being regulated by national and international standards led some scholars to defend the legitimacy of it because there is lawful and voluntary consent. However, autonomy can only be accepted on the question of organization and functioning of sports bodies if there is no break of the organization and operation. Only the law in the strict sense, emanating from Congress, can contemplate obligations. Anything else would violate national sovereignty. Questions may arise as to the legitimacy of specific regulations of FIFA on some topics, for example, the obligation to license agents only to individuals and the ban on clubs and coaches from using their services without a license (FIFA Players' Agents Regulations).[34]

FIFA has imposed Special Courts for trial on short-term sentences for cases of theft and robbery. This was seen during the South African World Cup. Thus, a private entity could end up meddling in matters unrelated to sport requiring courts to act in defense of purely commercial interests.

107. Make clear the limits that a private body (FIFA) has to establish standards, although the stated intent of it is to create a healthy environment in sports.

Explanation Employment contracts would considerably strengthen the position of professional football players with regard to their clubs by improving their legal position and affording better financial protection. Players have gained significant control from the clubs in recent times, but it is also true that agents ultimately control and potentially exploit players. Agents, not the authorities, are in control of a player's career destinations as they mediate contract negotiations and determine the player's choices.[35]

---

[34] South African authorities created 56 Special Courts for the South Africa World Cup (2010), which were imposed by FIFA, and arrested two Dutch women who wore miniskirts with the colors of a competing European Brewery that was not a FIFA sponsor. The women were released only after an out-of-court agreement. *See* FIFA exigiu mudanças na justiça da África do Sul. Jornal da Globo, June 23, 2010, http://g1.globo.com/jornal-da-globo. Accessed Jan. 28, 2014.

[35] Jonathan Magee. When is a Contract More than a Contract? Professional Football Contracts and the Pendulum of Power. 4 ESLJ xxvi, 2006–2007, page 46, April 1, 2013.

108. Give employee status to professional football players if it is not feasible to establish rigid controls of the activities of the player's agent.

Explanation Money diverted from illicit football takes away from investment in sport, from fans, athletes, potential athletes, etc.

109. Create an obligation to conduct an independent audit in sports federations and confederations.

Explanation Authorities herald to the world the importance of sport, which is considered a world good. Football, so-called "football-art," should not be subject to private appropriation before the manifest public interest. The consecration of the sport as a fundamental right obliges the government to visualize it, no longer under an economic prism and human relationships, but especially in social and cultural levels of paramount importance to the identity of people, which repudiate crime wherever manifestations happen. It is a matter of sport survival.

110. Treat football as a public good, particularly as a means of expression of different national cultures, and protect the historical and artistic bodies.

Bibliography

ABOVITZ, Ian. Why the United States Should Rethink Its Legal Approach to Internet Gambling: A Comparative Analysis of Regulatory Models that Have Been Successfully Implemented in Foreign Jurisdictions. 22 Temp. Int'l & Comp. L. J. 437, 2008.

BLANKENSHIP, Michael. The Unlawful Internet Gambling Enforcement Act: A Bad Gambling Act? You Betcha! 60 Rutgers L. Rev. 485, 2008.

BROWN, Charles E. Professional Football and the Antitrust Laws: Impact of *United States Football League V. National Football League* and a Proposal for Change. 31 St. Louis U.L.J. 1057 1986–1987, p. 1058.

DE SANCTIS, Fausto Martin. *Responsabilidade penal das Corporações e Criminalidade Moderna ou Responsabilida da pessoa jurídica e Criminalidade Moderna*. São Paulo: Saraiva, 2009

BANA, Anurag. Online Gambling: An Appreciation of Legal Issues. 12 Bus. L. Int'l 335, 2011.

EZABELLA, Felipe Legrazie. Agente FIFA e o Direito Civil Brasileiro. São Paulo: Quartier Latin, 2010, p. 41.

FEDERAL Bureau of Investigation—FBI, http://www.fbi.gov/about-us/investigate/vc_majortherfts/arttheft. Accessed May 21, 2012.

FATF, Vulnerabilities of Casinos and Gaming Sector, March 2009, available at www.fatf-gafi.org/dataoecd/47/49/42458373.pdf.

FIFA exigiu mudanças na justiça da África do Sul. Jornal da Globo, TV Globo June 23, 2010, http://g1.globo.com/jornal-da-globo. Accessed June 24, 2010.

FIFA denied Spanish Courts to get a copy of the contract celebrated between the Brazilian football player Neymar and Barcelona club (see http://esporte.uol.com.br/futebol/ultimas-noticias/2014/02/05/fifa-se-nega-a-enviar-contrato-de-neymar-para-justica-da-espanha.htm. Accessed Feb. 06, 2014).

GODDARD, Terry. *How to Fix a Broken Border: FOLLOW THE MONEY.*Part III of III.American Immigration Council Publication. Immigration Policy Center, May 2012.

GODOY, Arnaldo Sampaio de Moraes. *Direito tributário comparado e tratados internacionais fiscais*. Porto Alegre: Sergio Antonio Fabris, 2005, pp. 83–84

IRVING, James G. Red Card: The Battle Over European Football's Transfer System. 56 U. Miami L. Rev. 667, 2001–2002.

JENNINGS, Andrew. A ginga perfeita dos donos da bola. A FIFA controla o dinheiro, marca os adversários e dribla a Justiça. Entrevista. *O Estado de São Paulo*, J4, *aliás*, June 27, 2010.

JOY, Stephen; WARREN KINDT, John. Internet Gambling and the Destabilization of National and International Economies. Time for a Comprehensive ban on Gambling Over the World Wide Web. 80 Denv. U. L. Rev. 111, 2002–2003.

KARAQUILLO, Jean-Pierre. Droit International du Sport. 309 Recueil des cours 9, 2004.

KOBOR, Emery. *Money Laundering Trends. United States Attorney's Bulletin.* Washington, DC, vol. 55, no. 5, Sept 2007.

MACKENZIE, S.R.M. *Going, Going, Gone: Regulating the Market in Illicit Antiquities.*London: Institute of Art and Law—IAL, January 1, 2005.

MAGEE, Jonathan. When is a Contract More Than a Contract? Professional Football Contracts and the Pendulum of Power. 4 ESLJ xxvi, 2006–2007.

MARPAKWAR, Prafulla. State forms cells to detect source of terror funds. *Times of India.* Copyright 2011 Bennett, Coleman & Co. Ltd. December 24, 2011. www.westlaw.com. Accessed June 19, 2012.

MEXICO proposes to limit cash purchases of certain goods to 100,000 pesos. 2010 Fintrac Report.

MILLS, Jon. Internet Casinos: A Sure Bet for Money Laundering. 19 Dick. J. Int'l L. 77, 2000–2001.

MIRANDA RODRIGUES, Anabela. Contributo para a fundamentação de um discurso punitivo em matéria fiscal. *Direito Penal Económico e Europeu: textos doutrinários.*Coimbra: Coimbra ed., 1999.

OLSON, Eric L. *Considering New Strategies for Confronting Organized Crime in Mexico.*Washington, DC: Woodrow Wilson International Center for Scholars. Mexico Institute. Mar 2012.

ORMAND, Susan. Pending U.S. Legislation to Prohibit Offshore Internet Gambling May Proliferate Money Laundering. 10 Law & Bus. Rev. Am. 447, 2004.

PONS, Noel. Blanchiment d'argent: l'autre mercato. *Le Monde,*published on August 8, 2012. In www.lemonde.fr/sport/article/2012/08/30/blanchiment-d-argent-l-autre-mercato_1751790_3242.html. Accessed May 6, 2013.

SCHOPPER, Mark D. Internet Gambling, Electronic Cash & Money Laundering: The Unintended Consequences of a Monetary Control Scheme. 5 Chap. L. Rev. 303, 2002, Accessed April 1, 2013, p. 328.

SNAIL, Sizwe Lindelo. Online gambling in South Africa. Comparative perspectives. 15 Juta's Bus. L. 114, 2007.

SILVA SÁNCHEZ, Jesús-María. *Eficiência e direito penal.*Coleção Estudos de Direito Penal. São Paulo: Manole, 2004, No. 11, p. 11.

TERROR outfit-turned 'charity' JuD set o come under Pak Central Bank scanner. In *Asian News International*, March 13, 2012. www.lexis.com. Reporter, Naveed Butt.

THEODORO JÚNIOR, Humberto. *Consumers Rights (Direitos do Consumidor: a busca de um ponto de quilíbrio entre as garantias do Código de Defesa do Consumidor e os princípios gerais do direito civil e do direito processual civil).*Rio de Janeiro: Forense, 2009.

Index

F. M. De Sanctis, *Football, Gambling, and Money Laundering,*
DOI 10.1007/978-3-319-05609-8, © Springer International Publishing Switzerland 2014

Printed in Great Britain
by Amazon

29391729R00106